An ICT Primer

Information and Communication Technologies for Civil-Military Coordination in Disaster Relief and Stabilization and Reconstruction

Larry Wentz

Center for Technology and National Security Policy

National Defense University

July 2006

Larry Wentz is Mr. Larry Wentz is a Senior Research Fellow at the Center for Technology and National Security Policy, National Defense University, where he consults on Command and Control (C2) issues. He is an experienced manager, strategic planner, and C4ISR systems engineer with extensive experience in the areas of Nuclear C2, Continuity of Government C2, multinational military C2 and C3I systems interoperability, civil-military operations and information operations support to peace operations and numerous other military C4ISR activities. He is the author of *Lessons from Bosnia: The IFOR Experience* and *Lessons from Kosovo: The KFOR Experience.*

Mr. Wentz has extensive experience in business process reengineering, strategic planning and commercial telecommunications and information systems and their use in support of military C2. Prior to joining CTNSP, Mr. Wentz was a research scientist at the George Mason University Center of Excellence in C3I. Before his assignment at GMU, he completed an ASD C3I Command and Control Research Program (CCRP) sponsored lessons from Kosovo study. Mr. Wentz conducted a similar CCRP sponsored study of lessons from Bosnia in 1996-1998 when he was at ASD C3I as the Deputy Director of the Command and Control Research Program. Prior to the Kosovo assignment, he served as Vice President of Advanced Communication Systems, Washington and before that spent thirty years with the MITRE Corporation, a not-for-profit Federally Funded Research and Development Center—two-thirds of his MITRE career were spent in Europe supporting US/NATO C4ISR interoperability and the remainder as Technical Director of the Joint and Defense-wide Systems division of the MITRE Washington operation. Prior to joining MITRE, he spent 10 years with Bell Telephone Laboratories as a telecommunications systems engineer.

This paper is intended to be a living document. Please email comments, corrections, and clarifications to wentzl@ndu.edu. Updated versions of this paper will be available at http://www.ndu.edu/ctnsp/Defense_Tech_Papers.htm.

Defense & Technology Papers *are published by the National Defense University Center for Technology and National Security Policy, Fort Lesley J. McNair, Washington, DC. CTNSP publications are available online at http://www.ndu.edu/ctnsp/publications.html.*

Acknowledgements

This publication grew out of a study initiated by the Center for Technology and National Security Policy (CTNSP) at the National Defense University (NDU), in partnership with the Contingency Support and Migration Planning (CSMP) Directorate of the Office of the Assistant Secretary of Defense, Networks and Information Integration (OASD NII). The study was also supported by the Department of State's Office of the Coordinator of Reconstruction and Stabilization Operations (S/CRS) and Humanitarian Information Unit (HIU) and benefited from cooperation with Joint Forces Command (JFCOM) J9. Additionally, other organizations actively participated in and/or provided important insights and information. They included US government military and civilian elements, International Organizations (IO), Non-Governmental Organizations (NGO), UK civilians, "think tanks," academia and industry. Experts from these organizations lent their time, energy, wisdom and humor and provided invaluable insights from operational field experience, assessments of lessons learned, best practices, and other information products that helped make this primer possible.

In addition to the organizations already noted, other U.S. DOD military and civilian participants included the Joint Staff, National Geospatial Intelligence Agency (NGA), CENTCOM, SOCOM, Defense Information Systems Agency (DISA), US Army Civil Affairs and Psychological Command (USCAPOC), US Army TRADOC and 350th Civil Affairs Command, US Navy SPAWAR and NAVAIR, Pacific Disaster Center (PDC), and OUSD Stability Operations and Policy. Additional Non-DOD contributors included other State Department elements and the U.S. Agency for International Development (USAID). IOs such as the United Nations (UN) Department of Peacekeeping Operations (DPKO), the UN Office of the Coordinator for Humanitarian Assistance (OCHA), the World Food Program (WFP) and its Joint Logistics Center (UNJLC), the UN Development Program (UNDP), and the UN High Commissioner for Refugees (UNHCR) also participated in providing information. NGOs provided assistance, including Search for Common Ground (SFCG), Mercy Corps, Save the Children, DRASTIC/PACTEC, NetHope, Humaninet and Vietnam Veterans of America Foundation (VVAF). Civilian experts from the UK and the UK Post Conflict Reconstruction Unit (PCRU) also provided critical information, as did "think tanks" such as the Institute for Defense Analysis (IDA), US Institute of Peace (USIP) and Center for Strategic and International Studies (CSIS).

Numerous academic institutions also contributed their knowledge to this publication. These included the military Service schools and DOD institutions such as the Naval Post Graduate School (NPS), the NDU Institute for National Strategic Studies (INSS) including its Interagency, Transformation, Education and After Action Review (ITEA) Program, the NDU Industrial College of the Armed Forces, and the Army War College Department of Peacekeeping and Stability Operations Institute (PKSOI). Universities such as George Mason University, George Washington University and Eastern Mennonite University contributed as well. Finally, a variety of information and communications support to humanitarian assistance operations private industry players participated, including Groove, Global Relief Technologies (GRT), Morgan Franklin, Thoughtlink, and SAIC.

Contents

Foreword

The field of civil-military coordination in humanitarian disasters and post-conflict stabilization environments is characterized by rapid change and a pragmatic orientation toward real-life events—which occur much faster than the life-cycle of any publication. The temptation is to worry that any published attempt to provide a primer for professionals in this field may be overtaken by events as soon as the document goes to press. For that reason, this primer does not claim to be the last word on improving information and communications technologies (ICTs) for civil-military coordination. In fact, it does not even claim to be the last word of this project.

Much of this material was developed and refined through a series of workshops cosponsored by the National Defense University (NDU) and the Office of the Secretary of Defense, Networks and Information Integration (OSD NII). These workshops, held at NDU during 2005, drew participation and expertise from a wide variety of sources, within and beyond the U.S. government. The sponsors of this primer see no reason why this iterative process should stop, and every reason why it should continue.

For that reason, this primer should be considered Version 1.0. It is likely to generate further discussions, which will then be integrated with lessons learned and best practices from additional humanitarian disasters, stabilization and reconstruction (S&R) efforts, and complex emergencies. This is all to the good. As the universe of experience with civil-military coordination expands, further editions are likely.

The material generated for this publication should have a home beyond the written page. To allow full interaction with the text, and a fully transparent approach to its further development, an online version of the primer should be implemented, allowing additional postings and refinements by experts in the field. This electronic "living document"— available through a portal or hub website—would serve as the embodiment of the kind of information management best practices that it discusses. We have no doubt that this interactive approach is both necessary and highly desirable.

Introduction

The initial years of the 21st century have witnessed numerous large-scale crises, from the Indian Ocean tsunami and the Kashmir earthquake to longer-term, multi-faceted emergencies, such as those in Sudan. The United States has been involved as part of multinational coalitions in S&R missions in the Balkan states, Afghanistan, and Iraq. It has also provided humanitarian assistance in response to devastating natural disasters around the world. Increasingly, the scale and scope of such events involve both civilian and military components, as resources are stretched thin to support multiple ongoing crises.

Information and communications technologies (ICTs) are key elements of the global response to crises, whether natural or man-made disasters or post-conflict S&R scenarios. ICTs are vital enablers of the coordination mechanisms that civilian and military organizations need to assist local populations and host governments. ICT capabilities and requirements need to be better understood, so that relief and reconstruction efforts can be better constructed and coordinated by all parties working in the interest of the affected population.

This primer presents current knowledge and best practices in creating a collaborative, civil-military, information environment to support data collection, communications, collaboration, and information-sharing needs in disaster situations and complex emergencies. It consists of two parts. Part one defines and discusses the role of ICTs, the growing recognition of a need for improved collaboration, coordination, and information sharing, and the institutional and cultural characteristics of the various civilian and military participants in relief and reconstruction efforts. Part two draws real-world conclusions, provides best-practice recommendations, and offers practical checklists for maximizing use of communications and information management systems and processes.

The Universe of Crisis Response

As a threshold matter, it is important to address definitional issues involving different types of crises and the responses to those situations:

- *Humanitarian assistance and disaster relief* (HADR) operations generally follow rapid-onset natural or man-made disasters, such as hurricanes, earthquakes, or large-scale industrial accidents.

- *Stabilization and reconstruction* efforts often work in tandem with peacekeeping operations and are designed to stabilize and regenerate political and economic development in the aftermath of a conflict.

- *Complex emergencies* are often broadly scoped humanitarian crises that develop from the progressive and mutually reinforcing impacts of political and military conflicts;,economic collapses, natural disasters, such as famine or drought, and systemic problems, such as high rates of pre-existing poverty.

Clearly, these types of crises can differ in their causes and impacts, but the ICT support necessary for civil-military coordination in all of them has significant and numerous similarities, particularly in the earliest operational stages. These similarities stem from common occurrences in all kinds of natural and man-made disasters: significant (and perhaps wholesale) destruction of basic infrastructure and a breakdown in the societal mechanisms for individuals to obtain security, shelter, and other fundamental needs. Telecommunications systems are generally among the most vulnerable elements of infrastructure in any disaster or conflict, and their destruction or impairment can collapse a society's entire nervous system, threatening governance, economic activity, and basic security.

This primer is designed to be adaptable enough to assist ICT support professionals who are preparing or participating in future operations that might fit any one of the above definitions or, indeed (as many scenarios do) some combination of them.

The Case for Greater Cooperation, Collaboration, and Information Sharing

The purpose of this primer rests solidly on the premise that the growing complexity of crisis response efforts—and their increasing frequency and scale—call for a concerted effort to maximize the efficiency of communications and information management functions. Although many different groups and authorities can (and should) work in parallel, the pursuit of a common humanitarian purpose will be advanced through building what can be termed a *collaborative information environment* (CIE).

Constructing a CIE is not primarily a technology issue—effective, low-cost network equipment and data management systems exist today, and more are being developed. Rather, the challenges are largely social, institutional, cultural, and organizational. These impediments can limit and shape the willingness of civilian and military personnel and organizations to openly cooperate and share information and capabilities.

The recent spate of global disasters has sharpened the focus of attention on these socio-cultural issues and institutional disconnects. Many specialists in the humanitarian relief and reconstruction community are now striving to break through the organizational Gordian knot that often replicates itself in every new disaster, complex emergency, and post-conflict environment. Much work remains in developing an information-rich, reliable, replicable, and easily accessible humanitarian information space. This primer tries to distill the lessons that have been learned and provide tools to continue this work.

ICTs as Enablers

ICTs are increasingly recognized as enablers of greater coordination in crisis response operations and national capacity building. ICTs provide the means to link the constituent parts of an integrated response and the subsequent development and capacity-building efforts. Figure 1 illustrates how ICTs enable capacity-building initiatives in multiple sectors related to security, justice, governance, and economic and social wellbeing. For optimum effectiveness ICTs have to be given higher priority in planning and decisionmaking in terms of both investment and implementation.

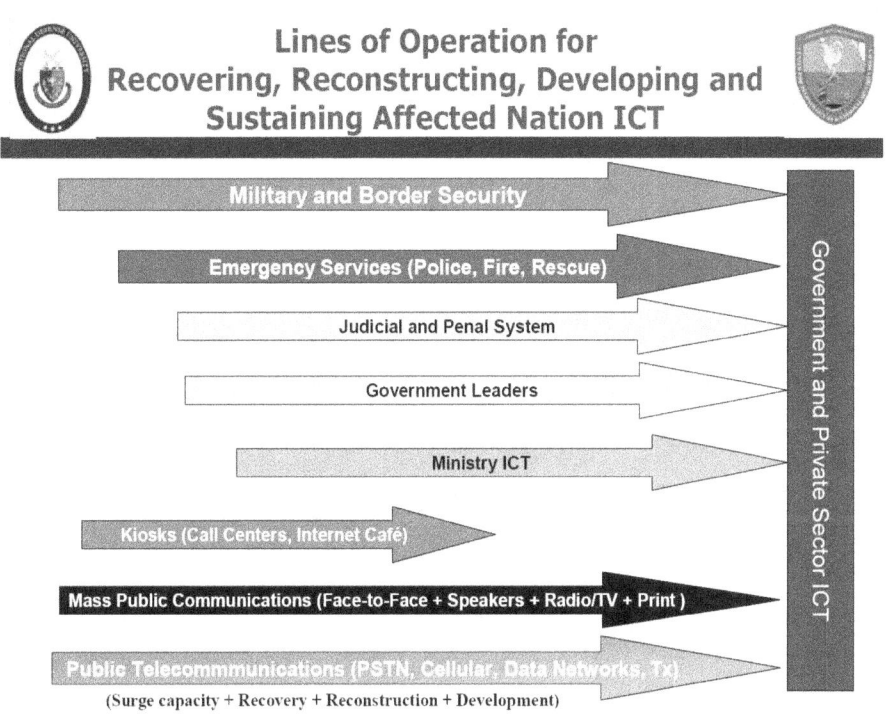

Figure 1

Several specialized non-governmental organizations (NGOs) and companies have recognized as a business opportunity the importance of rapidly installing ICTs in disaster response operations. Many now offer ICT packages that include commercial satellite communications capabilities, "Internet-in-a-box" products, turnkey and managed information services, and long-term, host-nation capacity building to support e-governance and other communication and information needs.

Part of the success of any effort to use ICTs to boost recovery efforts is what goes on before a disaster strikes or a complex emergency arises. Host governments, aid agencies, international organizations (IOs), and NGOs should be active in pre-planning and pre-positioning supplies and equipment, particularly in areas prone to natural disasters. Civilian and military responders alike need to take into consideration the need for coordination in their response plans. Because ICTs are critical enablers of relief and development efforts, there is a real need to understand and share information on the ICT infrastructure in any given country that could potentially suffer from a natural disaster or complex emergency. This baseline knowledge can then inform pre-planning for how to replace or augment vital ICT capabilities when a disaster does occur. Moreover, in the vital early days of a disaster response, an accurate baseline picture will aid in making a thorough assessment of where damage actually has occurred and will help identify and shape early ICT surge, recovery, and reconstruction initiatives.

Moreover, all stakeholders should keep in mind the long-range need to help host governments rebuild their ICT infrastructure. This means that, whenever possible, ICT actions taken in initial recovery stages should be done with a view toward leaving capabilities behind to jump-start the affected nation's ICT development and capacity building. Hence, near-term recovery and mid-term reconstruction improvements need to be consistent and synchronized with the long-term development objectives for host-nation ICT capacity building.

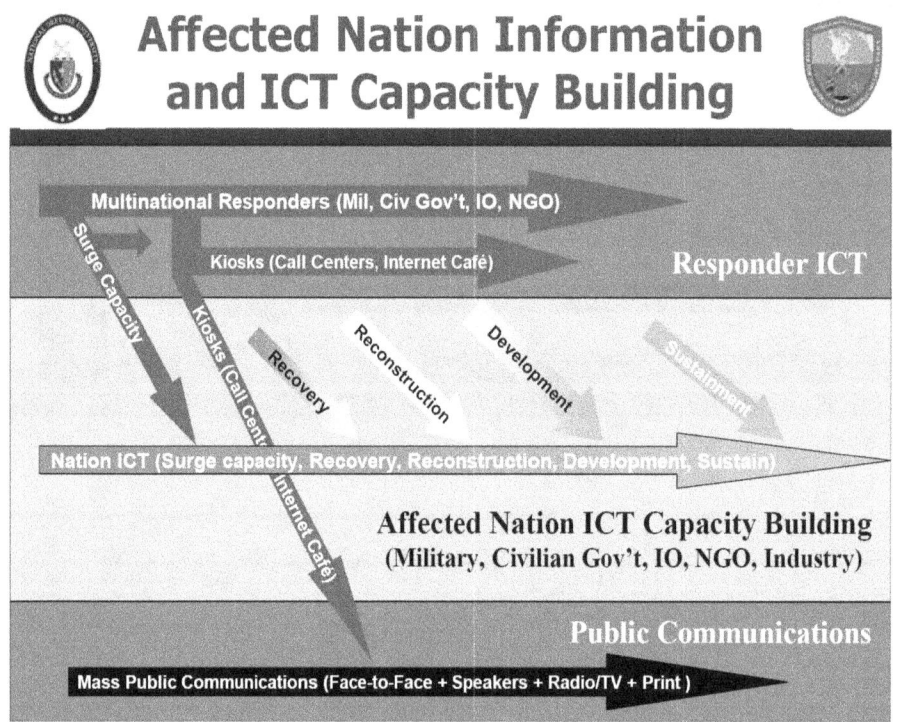

Figure 2

Figure 2 illustrates the need to incorporate public communications strategies into the planning and implementation of disaster responses. Disaster relief responders must be able to communicate with the affected nation's leadership and population. They can do this through phone kiosks, "tele-centers," and Internet cafes, but also through mass media such as radio, TV, and print. These capabilities need to be an integral part of the responders' ICT capability packages.

The Need for Civil-Military Coordination

Civil-military coordination is of critical importance in both the planning and execution of the recovery process, whether the risk stems from a hurricane or a failed state. To a large extent, there can be no development without security, nor can there be any real security without development—particularly over the long term, in areas where devastation has degraded socio-economic conditions.

Critical areas for civil-military coordination are: security, including rule of law; essential services, such as food, water, power, sanitation, medical, and shelter; logistics; communications; transportation; and information. The need for ICT capability is universal across these areas, and the way in which this need is met in the early stages of a disaster or complex emergency will be a key enabler in the host nation's ongoing capacity building. It is, therefore, important to give early attention and priority to ICT deployment.

The components of civil-military coordination consist of information and task sharing and joint planning—all of which depend on communications and management of data and information. The following issues, however, often complicate effective civilian-military coordination:

- A lack of understanding about the information culture of the affected nation;
- Suspicions regarding the balance between information sharing and intelligence gathering;
- Tensions between military needs for classification (secrecy) of data, versus the civilian need for transparency;
- Differences in the command and control style of military operations versus civilian activities; and
- The compatibility and interoperability of planning tools, processes, and civil-military organization cultures.

The sharing of information is particularly critical because no single responding entity—whether it is an NGO, IO, assisting country government or host government—can be the source of all of the required data and information. Making critical information widely available to multiple responding civilian and military elements not only reduces duplication of effort, but also enhances coordination and provides a common knowledge base so that critical information can be pooled, analyzed, compared, contrasted, validated, and reconciled. Civil-military collaboration networks need to be designed to dismantle traditional institutional stovepipes and facilitate the sharing of information among civilian and military organizations.

Experiences and lessons from real-world relief efforts and post-conflict recovery operations suggest the need to create a common culture of trust in information networks and communications between civilian governments, military organizations, IOs, and NGOs. Communications must flow in all directions, all the time. Information structures need to be flexible (but not ad hoc). Finally, lessons learned need to be understood after each events, and the required improvements must be institutionalized. All of this must happen in the context of multiple levels of interaction:

- Within organizations (including reach-back to home offices);
- Between organizations (bilaterally);
- Among organizations (multilaterally, as in a networked community);
- With local leaders and decisionmakers;
- With the media;
- Among the parties in any ongoing conflict; and, perhaps most importantly,
- With the local population.[1]

The global revolution in commercial ICT markets has contributed many valuable tools and removed many barriers to technical interoperability. But technology is only an enabler, trust is what sustains coalitions. Most of the remaining challenges involve reducing institutional, cultural, and social barriers that impede the building of trust and erode effective and timely collaboration.

The Role of DOD

Over the last several years, the Department of Defense (DOD) has become increasingly interested in the challenges associated with the transition from open conflict to post-conflict environments—to stabilization and reconstruction. DOD focus on S&R operations reflects the Nation's experiences in recent decades in responding to the complex suite of interrelated military, civil, economic, and political circumstances that arise in host nations and often influence the success of DOD missions in those areas of operation.

In addition, the U.S. military has been increasingly active in providing support services in global humanitarian responses to major disasters, including the 2004 Indian Ocean tsunami and Hurricane Katrina in 2005. There is a growing recognition that military forces have significant assets—including ICT capabilities—that can be brought to bear rapidly to assist civil disaster response, recovery, and early reconstruction efforts.

To assist in preparing ICT strategies for future S&R and HADR operations, DOD created the Contingency Support and Migration Planning (CSMP) Directorate within the Office of the Assistant Secretary of Defense, Networks and Information Integration (OASD NII). In conjunction with CSMP, the Center for Technology and National Security Policy (CTNSP) at NDU initiated a cooperative study of ICT support for S&R operations that became the basis for this primer.

DOD operates in tandem with, and often in support of, the civilian responder elements of the U.S. government, such as the Department of State (DOS) and the U.S. Agency for International Development (USAID), as well as IOs, such as the various United Nations

---

[1] Sheryl Brown, United States Institute for Peace (USIP) Virtual Diplomacy Series, "Creating a Common Communications Culture: Interoperability in Crisis Management," August 2005.

(UN) agencies and the International Federation of Red Cross and Red Crescent Societies (IFRC), and NGOs, such as Mercy Corps, OXFAM, and Doctors without Borders. Moreover, in any given deployment, DOD must work with and support the affected host-nation government and civil society.

A number of efforts are ongoing to strengthen U.S. government defense, diplomacy, and development initiatives to better link and coordinate responses to HADR and S&R operations. DOS has established the Office of the Coordinator for Reconstruction and Stabilization (S/CRS) to lead, coordinate, and institutionalize civilian government capacities to prevent or prepare for post-conflict situations and to help stabilize and reconstruct societies in transition from conflict or civil strife. U.S. military and USAID staff members have been detailed to S/CRS, which in turn participates in the exercises and other activities of the military's combatant commands (COCOMs). S/CRS is also reaching out to other governments, IOs, and NGOs.

Meanwhile, USAID has created an Office of Military Affairs (OMA), which serves as the day-to-day operational link with DOD and is responsible for the development of long-term strategic relationships. USAID is also strengthening its relationship with DOS, which works with USAID in dealing with other governments and IOs.

DOD has pioneered several initiatives, including the Afghan Reachback Office, which facilitates collaboration and coordination among DOD, DOS, and USAID for Afghanistan reconstruction activities. The COCOMs and other military elements, such as Civil Affairs (CA) units, are developing improved linkages with DOS and USAID. NDU and the military service schools and colleges are focusing on S&R operational training and education, including interagency coordination and outreach to the IO and NGO communities. The military educational institutions involved in such efforts include the Naval Post Graduate School Center for S&R Studies and the Army War College Peacekeeping and Stability Operations Institute.

In keeping with this ongoing process of developing and reinforcing interagency cooperation, this primer is an attempt to:

- Improve the ICT capacity of all players through appropriate civil-military collaboration, cooperation, and information sharing;
- Allow the U.S. government's ICT capacities—including those of DOD—to work more effectively with other players;
- Work with other nations' military and civilian government agencies—and with NGOs, IOs, and host countries—to facilitate civil-military support activities and enable the restoration, reconstruction, and development of host-nation ICT capacity;
- Develop a more informed understanding of roles, capabilities, and information needs among civilian and military elements of S&R and HADR operations; and
- Offer conceptual "toolkits" and best-practice recommendations for the use of ICTs in S&R operations, HADR relief efforts, and complex emergencies.

It should be noted that this primer, in its published form, is only one potential vehicle for discussion and dissemination of best practices, lessons, and suggestions from all parties in the field of ICT support for military-civilian coordination. Future efforts may well include the building of an online, interactive version of this document—a "living primer"—that will integrate the lessons gleaned from ongoing events as they occur. The goal is to provide a focal point for the sharing of knowledge and the airing of views, building a common body of knowledge and setting a foundation for mutual reinforcement of best practices—all in the interest of helping those individuals and societies, as yet unknown, affected by future disasters, conflicts, and complex emergencies.

PART ONE – THE NATURE OF THE CHALLENGE

Participants in Civil-Military Coordination[2]

The HADR and S&R operational landscape includes numerous and diverse participants, all with good intentions. Typically, as noted by Humaninet,[3] the participants "on the ground" consist of organizations and teams representing the following:

- Military forces of different nations;
- Developed-country government aid agencies;
- UN specialized agencies;
- Non-UN IOs;
- International and local NGOs;
- Host-nation governments (national, regional, and local);
- Volunteer, university, and faith-based teams and individuals;
- Corporate and business sector teams and assets; and
- Service providers and contractors.

The diversity of the responding organizations and teams is, in many ways, a real asset. It allows different organizations to bring to the table complementary capabilities, resources, and expertise. But the various groups also bring with them different agendas, operating principles, sensitivities, expectations, accountability mechanisms, and lines of authority. Often, military and civilian authorities have little control over many of the participants.

The information needed by each organization and team differs according to its mission, the situation on the ground, the needs of the population in the afflicted zone, the decisions of the host-nation government, and the phase of the response. It is a highly dynamic information environment. Informational needs can change by the hour and sometimes by the minute. Collaboration does happen, but often in an unplanned, unrehearsed way. HADR and S&R operations are necessarily messy and ad hoc, but the result is often a general lack of mutual understanding of roles, relationships, and capabilities. Not surprisingly, this contributes to a problematic lack of communication and information sharing among the civil and military elements of groups operating side by side.

Absence of trust is a fundamental source of tension among the civilian and military participants, as well as the local population and elites. Therefore, the key to success and relationship-building among individuals and organizations lies in understanding the roles, relationships, capabilities, motivations, and information-sharing needs in this complex environment. It is also vital to manage expectations and ensure that all actions support these expectations. Early introduction of ICTs to facilitate collaboration, cooperation, and information sharing among the civilian and military elements is important not only to the success of security aspects but also to the successful capacity building needed to restore

---

[2] Material in this section was derived from the Worldwide Humanitarian Assistance Logistics System Handbook provided by IDA for DOD, April 2004.

[3] HumaniNet, www.humaninet.org, a nonprofit organization that supports humanitarian relief teams with practical assistance in global ICT.

the host nation's governance and infrastructure and to begin the economic and social reconstruction process.

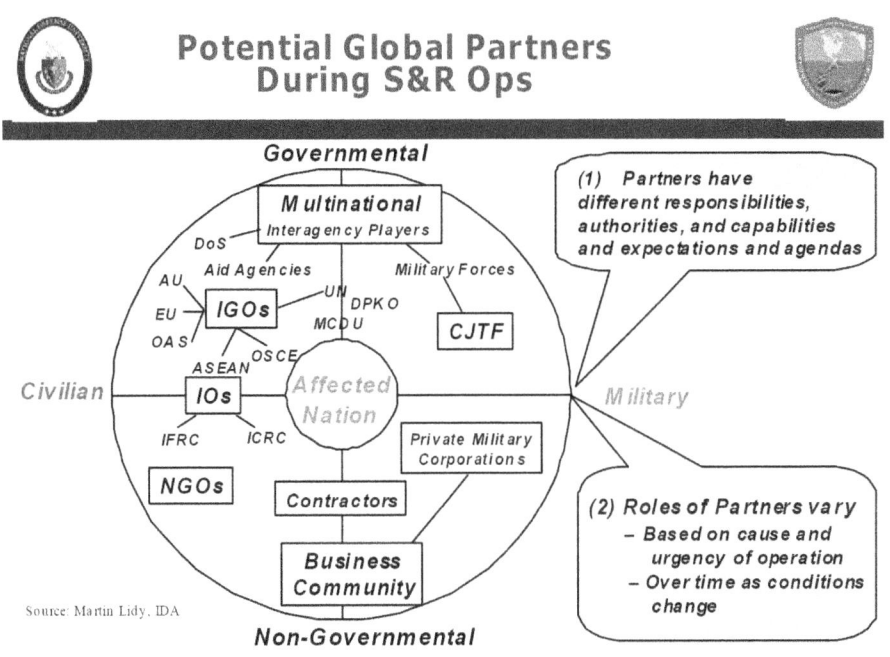

Figure 3

The Taxonomy of Participation

Figure 3 highlights the varied and evolving universe of agencies, institutions, and groups that provide different aspects of direct and indirect support. Still it is possible to define certain categories. Beginning with the U.S. government, as an example, the agencies likely to participate in domestic or international HADR and S&R operations include (but are not limited to) the:

- National Security Council,
- Secretary of Defense,
- Various DOD agencies,
- DOS (Coordinator for Reconstruction and Stabilization, Humanitarian Information Unit, and Regional Bureaus),
- USAID (the Office of Foreign Disaster Assistance, Office of Transition Initiatives, Field Missions, and Regional Bureaus),
- Department of Homeland Security (Federal Emergency Management Agency),
- Department of Transportation,
- U.S. Coast Guard,
- Department of Agriculture,

11

- Department of Justice,
- Public Health Service, and
- Immigration and Naturalization Service.

Other national governments have similar military and government agencies, such as in the United Kingdom the Department for International Development (DFID) and the Post-Conflict Reconstruction Unit (PCRU). Recognizing a gap in S&R response capabilities, Denmark recently has launched an initiative, the Concerted Planning and Action (CPA) of Civil and Military Activities, a cross-government committee to coordinate national responses to post-conflict operations.

Sovereign nations often provide significant humanitarian assistance to other nations, bilaterally or multilaterally, to help with development or to mitigate and manage complex emergencies and disasters. Most donor nations have created governmental departments or agencies to implement this assistance in accordance with their national foreign policy strategies and objectives. Typically, these governmental entities are responsible for long-term sustainable development in various countries throughout the world, and for providing rapid assistance when a disaster or emergency occurs.

To respond more effectively when time is critical, many of these agencies have formed organizations dedicated to disaster response. Some of these organizations provide teams, expert personnel, equipment, and supplies for humanitarian relief, whereas others contract significant portions of support to other government organizations or NGOs. Contributions may include in-kind or financial support, or a combination of both. In some cases, governmental bodies, such as civil defense organizations and aid agencies, may provide rapid response from standby personnel in response to appeals from affected nations' governments or humanitarian agencies like those of the UN. Typically, these agencies divide the globe into geographic regions and assign areas of responsibility (AORs) to functional and regional bureaus.

As representatives of their governments, overseas offices of aid agencies are often located at, or very near, their nation's foreign embassies or consulates, and the assistance they provide is typically offered without expectation of reimbursement from the affected nation. Aside from USAID and DFID, which already have been mentioned, examples of donor agencies include the Australian Agency for International Development (AUSAID), the Swiss Agency for Development and Cooperation (SDC), and the Canadian International Development Agency (CIDA).

IOs such as the International Committee of the Red Cross (ICRC), the IFRC, and the Red Cross and Red Crescent National Societies are neutral intermediaries that play a significant role in supporting HADR and S&R operations. The ICRC is a private Swiss institution that acts as the "guardian" of the Geneva Conventions and intervenes during armed conflicts to protect and assist victims and visit prisoners of war and civilian detainees. It also participates in relief programs for displaced persons. The IFRC, on the other hand, is a coalition of national societies that intervenes to coordinate relief efforts for victims of natural disasters. In the event of armed conflict, it provides humanitarian assistance to displaced individuals, in coordination with the ICRC and the national societies.

More than 180 nations have established their own national societies to offer humanitarian assistance and relief after armed conflicts and natural disasters. In most countries, these societies are known as the "Red Cross," but in many Muslim countries they are known as the "Red Crescent." In Israel, the national society is called the Magen David Adom. These Red Cross/Crescent organizations are guided by seven principles: concern for humanity, impartiality, neutrality, independence, voluntary service, unity, and universality. Humanity, impartiality, and neutrality also serve as the underlying principles of international humanitarian law and the Geneva Conventions.

Inter-governmental organizations (IGOs) also cooperate on specific issues, such as economics, security, culture, politics, or common geographic concerns. IGOs promote common policies and preventive diplomacy, and they seek to implement international agreements, resolve disputes, and foster collective security. Some organizations, such as the UN, are globally focused and draw their membership from more than one region, culture, or language. IGOs such as the International Organization for Migration (IOM) are functionally based and include a wide range of member states and partners focusing on a particular issue.

The UN system is a complex group of organizations, at the center of which is the UN Organization itself, which is composed of member states around the world. The member states make up the General Assembly and serve on the Security Council and various committees, all of which are supported by the UN Secretariat working out of New York. In terms of peacekeeping operations, the UN Department of Peace Keeping Operations (DPKO) takes the lead in putting together and managing UN peace operations. In such operations, the UN will appoint a Special Representative of the Secretary General (SRSG) to represent the UN on the ground and to manage military and civilian political operations. [4]

In terms of humanitarian response, the UN coordinates the efforts of its various agencies through its Office for the Coordination of Humanitarian Affairs (OCHA). In countries affected by humanitarian emergencies, the UN will normally appoint a Humanitarian Coordinator to take the lead. The vast majority of UN humanitarian activity, however, is undertaken by the separate UN agencies, which include the:

- World Food Programme (WFP);
- UN High Commissioner for Refugees (UNHCR);
- UN Children's Fund (UNICEF);
- World Health Organisation (WHO); and
- UN Development Programme (UNDP).

---

[4] Based on writings of Paul Currion, a consultant on Information Management for Humanitarian Operations. See http://www.currion.net/inno.htm.

These UN agencies are autonomous from the UN central organization and maintain their own management and administrative structures. Each UN agency undertakes a wide range of activities, including food distribution, infrastructure rehabilitation, health provision, water and sanitation, education, and protection of populations. In many humanitarian operations, one particular UN agency will be named the "lead agency" for coordinating a particular sector, and UN agencies will frequently sub-contract with NGOs to carry out specific tasks.

Other IGOs retain a regional focus in dealing with a multitude of issues. For example, the North Atlantic Treaty Organization (NATO) is a political/military organization focused on the security and defense of Europe, North America, and the Atlantic Ocean basin. Since the end of the Cold War, NATO has also assumed an extensive role in UN Security Council-authorized regional peace support operations. The Organization for Security and Cooperation in Europe (OSCE) is a regional organization that fosters security and economic cooperation through the implementation of human rights, fundamental freedoms, democracy, and the rule of law. It brings together all the countries of Europe, Canada, and the United States as well as certain countries in Central Asia.

NGOs are of many different types and sizes, ranging from major global organizations such as CARE and World Vision, to small charities. NGOs are established for many purposes and include political parties, labor unions, religious and social groups, community interest groups, research institutes, churches, professional associations, and lobbying groups. In humanitarian operations, international and national NGOs are important actors, channelling large amounts of private and public funding. NGOs carry out a wide range of activities in a variety of ways. Many of them have a sectoral focus, such as water and sanitation; a thematic focus, such as children or refugees; or a regional focus, such as Africa. Some NGOs are operational and carry out physical activities in the field while others focus more on policy or advocacy, for example, Amnesty International. It is impossible to list the complete range of activities that NGOs undertake—one organization known as Clowns Without Borders provides educational entertainment to children affected by disasters. But in general, NGOs work to eliminate the causes of humanitarian emergencies and mitigate their effects on human populations. Some of the larger, more experienced NGOs that respond to disasters also conduct long-term development projects.

NGOs do not have the same legal sovereignty that IOs enjoy because most are not formed under international law. They are subject to the laws of the affected nations in which they work. They can support governmental entities through contracts or grants, but otherwise they typically have no formal authority and tend to act independently.[5] NGOs are, however, responsible to their boards of directors, accountable to their private contributors, and responsive to the particular governments that chartered them as nonprofit entities. Although they may take money from one or more government sources, NGOs are not instruments of their governments and do not usually take policy direction from institutional donors. Additionally, the UN has no mandate to direct the work of NGOs, nor do representatives of any governmental or intergovernmental organization.

---

[5] NGO is the equivalent of the term private voluntary organization (PVO), sometimes used by U.S. groups.

However, the UN does have a role to play in coordinating overall humanitarian operations, which may include NGOs within the "lead agency" system.

The following list represents a sampling of the non-government community of participants: the IOs, IGOs, and NGOs. While it is by no means all-inclusive, it provides some sense of the staggering array of groups that may show up during a world crisis:

- IOs
  - ICRC
  - IFRC
  - National Red Cross and Red Crescent Societies, for example the American Red Cross
  - SMOM (Sovereign Military Order of Malta)

- Inter-Governmental Organizations (IGO)
  - UN agencies and offices
    - Office for the Coordination of Humanitarian Affairs (OCHA)
    - Department of Peacekeeping Operations (UNDPKO)
    - UNHCR
    - WFP
    - UNICEF
    - Food and Agriculture Office (FAO)
    - WHO
    - UNDP
  - World Bank Group
  - International Monetary Fund (IMF)
  - NATO—Formally Linked to UN
    - Civil Emergency Planning
      - Euro-Atlantic Disaster Response Coordination Center (EADRCC)
      - Euro-Atlantic Disaster Response Unit (EADRU)
      - NATO Air Traffic Management Committee (NATMC)
    - Strategic Commands
      - Supreme Headquarters Allied Powers Europe (SHAPE)
        - Combined Joint Task Force (CJTF), such as IFOR, SFOR, KFOR, ISAF
        - Civil-Military Coordination (CIMIC)
      - Allied Command Transformation (ACT)
    - Partnership for Peace (PfP) Nations

- IGO—Functional/Regional Focus
  - International Organization for Migration (IOM)
  - European Union (EU)
  - European Community Humanitarian Aid Office (ECHO)
  - OSCE
  - Caribbean Community and Common Market (CRICOM)

- o Organization of American States (OAS)
  - o Coordination Center for the Prevention of Natural Disaster in Central America (CEPREDENAC)
  - o African Union (AU)
  - o Economic Community of West African States (ECOWAS)
  - o Association of Southeast Asian Nations (ASEAN)

- NGO
  - o Cooperative Assistance for Relief Everywhere (CARE)
  - o World Vision
  - o Doctors Without Borders (Médecins Sans Frontières-MSF)
  - o OXFAM
  - o Church World Services (CWS)
  - o United Methodist Committee on Relief (UMCOR)
  - o Mercy Corps International (MCI)
  - o International Rescue Committee (IRC)
  - o Catholic Relief Service (CRS)
  - o International Medical Corps (IMC)
  - o Danish Relief Council (DRC)
  - o Norwegian Relief Council (NRC)
  - o Save the Children
  - o Alliances of NGOs:
    - InterAction (U.S.-Based NGOs)
    - International Council of Voluntary Agencies (ICVA)

Many of these organizations, such as the UN, CARE, Mercy Corps, and MSF often are already operating in a crisis area when the military components arrive, and they continue to do so, adapting their services to the contingencies that arise in a crisis. Figure 4 compares selected characteristics of the potential participants.

Characteristics of Potential S&R Ops Global Participants

Characteristics	IO	IGO	NGO	Business
Formed for a specific purpose	X	X	X	X
Consultative body of National Governments		X		
Formed under International Humanitarian Law or Custom and Recognized as a sovereign entity	X			
Directed by representatives of National Governments		X		
Directed by private citizens	X		X	X
Funded by National Governments	X	X	X	X
Funded by private institutions or individuals	X		X	X
Not for profit entity	X	X	X	
For profit entity				X

Research documented in IDA Document D-2349 "Potential Global Partners in Complex Contingencies," currently in working draft.

Figure 4

The international business community is also a key participant in HADR and S&R operations. Corporations have extensive capabilities to generate employment, investment, and economic growth in much of the world, and they maintain significant resources that are useful in foreign humanitarian relief and development operations. Many different types of businesses become involved in emergency operations, and for very different reasons. When a disaster or emergency occurs, companies are in a position to provide technical expertise, donate their products, or contribute financially to humanitarian response organizations. In potentially volatile areas, multinational corporations with a stake in a country or region have a vested interest in promoting stability and preventing conflict situations that could disrupt their business. Multinational corporations are often very knowledgeable about the host nations in which they operate. They establish working relationships with local and national authorities, employ local personnel, and have an understanding of the local dynamics and available resources. Moreover, they have fully developed supply chains.

Additionally, the use of private military companies (PMCs) has emerged as a growing and controversial addition to the increasingly complex operations landscape, particularly in S&R or complex emergency scenarios. These companies provide services that until recently were considered the exclusive domain of national armed forces. Their customers include emerging national governments and military establishments, multi-national corporations and—increasingly—civilian and humanitarian relief organizations. Their presence is not universally understood, appreciated, or accepted; they are often referred to as "mercenaries." Just as often, misunderstandings within the PMC industry itself give credence to this charge and do little to increase understanding. A legitimate PMC is not a mercenary organization. The PMC trend presents significant opportunities and economies to governments and humanitarian and civilian agencies alike. On the other hand, it can present serious risks to mission success for these same customers. It is absolutely essential for leaders within the military and civilian communities to understand both the risks and opportunities PMCs represent. A sample of companies falling into this category would include MPRI, DYNCORP, KBR, CACI, and BAH.

Levels of Internal Support

The U.S. military deploys with a comparatively high standard of internal logistical and material support, and in fact, the number of personnel, support packages, and overall "baggage" may seem excessive to others. But this support is designed to make the military as self-sustaining and self-reliant as possible. The practical value of this support apparatus is that it can be used to assist displaced civilians. However, the military standard of support is based on national government policy determinations and may differ from humanitarian relief agency standards. In addition, the military will constantly try to avoid "mission creep," which occurs when armed forces take on broader or additional missions than those for which they initially planned or trained.

Meanwhile, the mandate for internal force protection overrides nearly every other concern for the military. The term *force protection* refers to security procedures and arrangements to protect soldiers, civilian employees, facilities, and equipment that are part of the military organization. How force protection is implemented may have an effect on how, where, and when the military will become involved in relief activities.

Each military operation will have *rules of engagement* (ROEs), which delineate the circumstances and limitations under which the military will initiate or continue combat. The ROEs also have significant impacts on disaster relief operations, affecting freedom of movement, security, logistics, and the perception of neutrality of the relief community in the eyes of competing factions. The military's purpose generally is to create a safe and secure environment to enable the civilian government, IOs, and NGOs to conduct their humanitarian assistance, reconstruction, and development missions. Secondarily, force protection also is designed to protect those military elements directly involved in transport or other support activities.

Information Needs[6]

Having understood the breadth of participation in HADR and S&R operations by multiple civilian and military actors, we now turn to the question of what particular information these entities need to fulfill their missions. The informational needs of military and humanitarian actors are different, even when they involve the same subject or topic. This stems from differences in for what purpose the data will be used and what implications the data may have for humanitarian operations. As a threshold matter, several principles should be kept in mind when discussing these information needs:

- Common terms used by the military and civilians often have different meanings. For example, to a military organization the term *sector* means a geographic area of responsibility, but to humanitarian organizations it denotes a functional area, such as water/sanitation, food, or shelter.

- The urgency of the need for humanitarian data may not always be readily evident to the military. When a data request is passed to the military, it may not always be treated with the highest priority, due to other informational needs deemed to be more pressing. It is best to avoid depending on the military for data that is time-sensitive and has not been shared before.

- When requesting data or information from the military, it is better to state the need in terms of the decision that must be made instead of the raw data that is being used to make that decision. For example, imagery of mountain passes is useful to humanitarians in determining when those passes might close to transportation. Instead of asking directly for imagery, it may be better to ask for a forecast of when the passes will close.

Performing a Needs Assessment

Organizations should perform a "needs assessment"[7] as part of their planning for what kinds of information services and products they will need, both before and during deployment. A needs assessment examines what kinds of information organizations and individuals will need to make effective decisions, and how they will receive (or not receive) that information. The aim of the assessment should be to get a clear picture of what is known as the *information environment*—the sources and markets for information, the flows that connect those sources and markets, and the institutions that mediate those flows. The basic questions that an information needs assessment should answer are:[8]

---

[6] Material in this section was derived from a presentation by Dennis King, U.S. DOS Humanitarian Information Unit, entitled "Humanitarian Knowledge Management," given at the second international ISCRAM conference, Brussels, Belgium, April 2005.
[7] Currion.
[8] UN OCHA Civil Military Coordinator Course training material distributed at Santo Domingo, Dominican Republic, May 15-20, 2005.

- What information is needed? Information needs cannot be determined in isolation from the rest of the disaster response. It is important to establish what activities the humanitarian community is planning and to broadly identify what sort of information will be needed to support these plans. Although there is a large body of core data needed in any emergency, differences in emphasis affect the collection and dissemination processes. Establishing information needs should be done in consultation with heads of agencies and any coordination bodies to ensure that the entire humanitarian community feels some ownership of the process and content.

- Who is going to use this information, and for what? The target audience and planned purpose of the information are both factors that will affect what methods should be used to collect, process, and present it. Be aware that the needs of NGO staff members who are implementing projects in the field are very different from the needs of heads of agencies.

- What information sources already exist? Before embarking on a costly information-gathering exercise, one should make every effort to verify that this information does not already exist. It may help to contact other organizations, both in the field and outside the country, which might already have relevant information and be prepared to share it.

- What resources exist for managing this information? Who is prepared to take on the costs of information management? If one can identify which organizations are going to be gathering information as a normal part of their activities, it is easier to support and improve their efforts. Information gathering in any response operation has five broad stages: collection, processing, analysis, dissemination, and incorporation of the new information into the decisionmaking process.

- What constraints are there on successful information management? Security concerns can sometimes limit the collection of data, particularly where access to specific areas or groups is dangerous. Similarly, sharing information contained in a report might be politically sensitive. Competition for funding also jeopardizes information sharing when organizations feel that others may use or misuse information they have collected. All of these concerns must be addressed.

Reliability Issues

Identifying information exchange needs is not easy. Natural disasters, humanitarian emergencies, and S&R operations are, by their very nature, complex and dynamic situations. They are multi-sectoral and multi-disciplinary, incorporating both the physical and social sciences. Information comes from a multitude of different sources and is constantly changing. It is often incomplete or contradictory. In some cases, there is an overload of information, and in other cases, there are complete gaps in what is known. Collecting information is often difficult, if not impossible, because of inaccessibility to the affected areas due to natural hazards, lack of a safe and secure environment, or government restrictions. Furthermore, much of the "available" data are actually estimations based on selective sampling or extrapolations of dated statistics, such as census information, projected growth rates, and proxy indicators.

Also, a certain amount of misinformation and disinformation can be generated during S&R, HADR, and complex emergency situations. Governments and aid organizations may publish inflated or distorted estimates or data to appeal for higher amounts of international donor assistance. Any participating civil or military element may purposely conceal information it deems sensitive. The end result is that decisions about providing assistance or support often must be based on the best available information, however incomplete and insufficient.

Another challenge is the inconsistent use of standardized meta-data when collecting and providing information. All incoming and outgoing data and information should include a type of provenance: the source and the date or time-stamp, so that other users can determine the credibility and currency of the content. It is also important to clearly define any ambiguous terminology and explain methodologies and indicators used to collect the data. Finally, data and information should be geo-referenced to include the latitude/longitude, geo-code, gazetteer place name, administrative unit and other identifiers. This allows the data to be entered into a *geographic information system* (GIS) and mapped. If these standards are followed, data and information provided by many different civilian and military organizations can be effectively pooled, compared, contrasted, validated, and used for analysis, mapping, and operational activities.

What Do Organizations Need to Know?

In any deployment, both civilian and military response organizations want answers to certain questions. For example, all organizations need background and situational information. But other types of needed information are more organization or specialist specific. Strategic-level policymakers, for example, want "big picture snapshot" analyses to understand the issues, make decisions on providing assistance, and identify problems and obstacles. Field personnel and project and desk officers, on the other hand, need more detailed operational and programmatic information to plan and implement humanitarian assistance and reconstruction programs. Most information needs can be divided into the following basic categories:

Background/baseline information: Relief agencies and organizations need information about the country's unique history, geography, population, political and economic structure, and culture. Beyond these general characteristics, they need more specific, baseline data reflecting the "normal" (that is, pre-disaster) conditions prevailing in various relevant sectors, such as, medical, transportation, communications, and food supply. Baseline data are also necessary so that relief organizations can compare the actual conditions they find during a complex emergency or following a natural disaster or conflict with prior conditions. Background and baseline information tells them the:

- Country's pre-crisis population (national, province/state, city/town) and its demographic composition (ethnicity, religion, age cohorts, urban/rural, political);
- Pre-disaster geography(especially relevant following earthquakes, flooding or volcanic eruptions);
- Recovery record from past disasters and natural hazards;
- Most recent annual baseline health indicators for the population (crude mortality rate, infant/child mortality rates, HIV adult prevalence, and malnutrition);
- Recent annual economic indicators, such as gross domestic product (GDP), gross national product (GNP), agricultural/food production, and staple food prices).

Situational awareness information: Once the crisis has developed, organizations need to know the latest about the situation on the ground and information about the conditions, needs, and locations of affected populations. Some examples of situational awareness questions are:

- What is the latest or most current humanitarian and reconstruction situation in the country?
- What are the most recent severity indicators (death tolls, mortality rates, malnutrition rates, economic impact, and infrastructure damage)?
- Who are the affected populations (refugees, children and other vulnerable groups, and resident populations), how many are there, and where are they located?
- What is the assessment of damage to infrastructure (transportation, buildings, housing, and communications)?
- What is the latest or most current security situation in the affected areas of the country?

Operational/programmatic information: Organizations need information to plan and implement humanitarian assistance and reconstruction programs. Questions that program planners may ask include:

- Where are (and what are the conditions of) the logistical access routes for delivering humanitarian and reconstruction assistance?
- What assistance organizations are working in the country, what are their programs and capacities, and where are they working?
- How is the host-country government responding and can it provide more assistance itself?
- What are the programmatic/financial needs of the responding organizations?
- What (and how much) funding is being provided to the response organizations and who are the donors?

Analysis information: Data must be compiled and put in context to become useful as information. Information, in turn, needs to be interpreted and analyzed in relation to other thematic information. Analysis can include evaluations of issues and responses, projections about the future, and recommendations for policy decisions and operational deployments. Analysts can be expected to ask:

- What were the causes and contributing factors of the emergency?
- Have any constraints arisen in providing humanitarian and reconstruction assistance in the crisis, such as insecurity, inaccessibility, government interference, or other problems?
- How effective are ongoing humanitarian and reconstruction assistance programs and responses?
- What are the short-, medium- and long-term impacts of the emergency?
- How are evolving political and economic conditions likely to affect assistance efforts in the future?
- What are the options and recommendations for action?

Where Can Information Be Obtained?

The emergence of the Internet in the last 15 years has revolutionized the availability and dissemination of humanitarian and S&R-related information. Email has greatly facilitated the transmission of information between the headquarters of the humanitarian and S&R response organizations and the personnel, teams, and programs located in the field. The World Wide Web provides a vast virtual library of information to users with Internet access.

At the same time, the Internet has added to the overload of information and the increasing difficulty in locating, extracting, and verifying the answers to critical questions. In fact, the Internet can complicate the tasks of information and knowledge management. Information assurance is just one of the many challenges—trust and self-policing are the norm for information sources on the Internet. No one organization is responsible for assuring the quality and integrity of information, and no set standards are in place for collecting and populating the various web sites with that information. All of this adds to the challenges of sharing information.

Situational awareness information often comes in a seemingly common-sense form: news reports. But better sources are often found in the situation reports and field assessments produced by the response organizations working in the affected countries. These organizations also draft and issue appeals, proposals, and project monitoring documents that provide operational and programmatic information. Useful background/baseline information can be found in country profiles, maps, databases, and chronologies. Analysis is derived also from evaluations, lessons learned, research studies, and policy recommendations.

Not everything that S&R and HADR organizations need to know, however, can be found in databases, documents, and visual products. Tacit knowledge is usually not documented, but rather derives from expertise, collaboration, and field experience. This knowledge is often imparted in briefings, discussions, and first-hand observations. "Seeing it for yourself" adds a great deal to one's knowledge and understanding of any humanitarian emergency.

Military Information Sources

A large number of sophisticated intelligence collection systems are available to military participants. Most are designed for collection of information relevant to military operations. Consequently, many of these systems are extremely classified, and the information collected by these sources is carefully controlled by the military. Most information relevant and available to humanitarian responders will be collected on the ground.

The military is more likely to be effective in gathering data about tangible, measurable things, such as the length of an airfield and the number of structures in a village with intact roofs. It seldom has the skills to make more than broad observations about issues such as food security, access for ethnic minorities, or other broad-scope, demographic measurements.

For many militaries deployed outside of their own countries, direct interaction with the population is limited by policy, regulation, or tactics. Effective interaction, when it is permitted, may be further limited by language, culture, or distrust. Lower-ranking soldiers often collect much of the information, and they may lack cross-cultural or language skills relevant to the host population. In some militaries, however, dedicated, trained personnel interact with the population and conduct CA patrols to collect or verify data. It is important to know how the data was collected and whether or not it has been verified before taking action.

Civil-Military Cultures and Challenges

Most organizations have a distinct organizational culture—an often-unwritten set of rules, regulations, viewpoints, perspectives, and operating procedures. This culture is based on the unique history, mission, structure, and leadership of the organization. This section explores how differences between military and non-military organizational cultures affect crisis response operations and military-civilian coordination.[9]

Military Organizational Culture

The military's distinct organizational culture, which makes the military very effective in combat, may frustrate civilian relief organization personnel, who may find the military inflexible and inscrutable. For example, military organizations generally are:

- Highly structured, hierarchical, and oriented toward "stove-piped" chains of command;
- Authoritarian and command-oriented;
- Focused on attaining mission goals (both explicit and implied);
- Bound by extensive rules and regulations;
- Characterized by elaborate process and scheduling regimes (the "daily battle rhythm");
- Driven by a "work hard, play hard" ethic;
- Highly competitive;
- Respectful of internal traditions;
- Respectful of physical and mental toughness;
- Respectful of age, experience, and seniority;
- Trained toward an ideal of combat readiness;
- Trained for battle skills, physical fitness, and equipment maintenance;
- Trained to place a high priority on their own battlefield survival;
- Trained to be secretive for operational security;
- Led by officers who are taught to be assertive, decisive, tenacious, and confident, and to "make a decision and make it now!"
- Apt to avoid or discount values of cooperation, collaboration, and nonconformity.

By contrast, civilian relief organizations often have radically different organizational cultures and structures. They tend to be less formal, less authoritarian, and less focused on internal traditions or security concerns. NGOs, for example, have a much "flatter" structure than the military. Normally, an NGO is headed by an executive officer (no standard naming convention exists, but the top official may be called a Country Director, Head of Mission, Chief of Party, or Program Manager). Underneath the executive officers, commonly will be a senior staff, which may include project managers, administrative officers, and security officers, among others. The next layer of management may be the project staff—water and sanitation engineers, medical staff,

---

[9]The following discussion draws heavily on the UN OCHA Civil Military Coordination Course.

media officers, and others. Often a mix of international and national staff members work side by side within NGOs, particularly at the project level. Because of this flatter structure, most NGO staffs have a greater level of flexibility and autonomy than their military counterparts.

Some NGOs are reluctant to collaborate with military entities and some are not. It depends on the NGO, the individual staff members representing that NGO in the field, and the situation in the country. If the military units are part of a UN peace operation, then NGOs are more likely to be receptive to cooperation. However, if the military units are belligerents in a conflict, NGOs are unlikely to want any working relationship with them. The reason for this is fairly simple. To maintain their neutrality and impartiality, NGOs cannot be seen as favoring or supporting a specific side in a conflict. Even if the NGOs' principles are not being compromised through cooperation with belligerents, they must be sensitive to the perceptions of the local communities and authorities they seek to help. This is also the reason that many NGOs, traditionally, do not use armed guards for their staff or premises, as this might undermine their acceptance with the community.

Moreover, most NGOs simply do not have much experience in working with the military. The UN and NGO communities are in the process of developing clearer policies on civil-military relations. But while these are developed and implemented, all sides must continue to work with each other in the field—often overcoming an acute lack of mutual understanding of each other's goals, priorities, or processes.

Both military and civilian organizations are diversified on a functional basis with differing organizational structures. On the military side, the administrative, intelligence, operations, logistics, and CA elements report directly to the commander. In civilian organizations, the functions of the administrative, public information, program, supply, transport, and protection units are more autonomous. In field operations, this difference can prove frustrating for both the military and the civilians. Heads of humanitarian offices are not commanders in the military sense. They may lack the single point of contact the military seek. This means, for example, that a military commander may have to deal with a logistics manager on operational matters and a different manager on legal matters. Without knowledge of particular civilian organizations' functional divisions, this can be confusing to the military.

Another common area of difficulty is the differences in logistical support networks. Civilian and humanitarian organizations often lack rear support. NGO operations are almost entirely donor-driven and subject to "pipeline" delays. Much of their decisionmaking authority devolves to staff members in the field. Moreover, civilian "command and control" (C2) does not always emanate from a single central point.

Counterparts to military officers—and especially NGO counterparts to military officers—tend to receive decisionmaking authority at a younger age than do military decisionmakers. This generation gap has been known to exacerbate the cultural difference between civilians and the military.

Civilian organizations commonly seek to achieve maximum efficiency in the use of limited resources. They stretch material and human resources as far as possible and tend to be less geared toward end states or goals. Accountability may be less rigorous than in the military. Civilian organizations tend to pride themselves on being light-footed and flexible in emergency situations, yet may lack a facility for long-range planning. Hence, civilian flexibility and military precision may conflict in joint operations. Civilians may be inclined to judge military achievements solely by their end results without paying due regard to the careful training, preparation, and planning upon which military success is founded. Hence, outsiders may perceive military units as well-resourced in human and material terms and then form unrealistic expectations of what those units will do.

Conversely, military personnel often expect field meetings to be highly structured and efficiently managed. A military leader is expected to listen to succinct briefings (usually accompanied by PowerPoint slides) and make clear decisions without hesitation. Military representatives will come to meetings expecting everyone to leave with their "marching orders." As a result, they may view civilian meetings that wind their way slowly toward a consensus as direction-less, inefficient, and lacking in leadership. Sensing what they perceive as a "vacuum" in direction, military officers may attempt to assert their own leadership—intending to help. Other military personnel may simply become frustrated, lose interest, and not participate further.

The military is highly concerned about operational security, which results in a reluctance to share information about planned military activities. This does not keep the military, however, from wanting in-depth information about civilian activities. The military will respond well to clearly stated missions, efficient processes, organization, responsibility, and competence. It will judge harshly any operation weak in these areas and may show insensitivity when expressing that judgment.

Some military leaders may be concerned that humanitarian operations degrade combat readiness. Sensitivity to suffering is not always viewed as a virtue on the battlefield. This can result in a desire to minimize participation in some operations. Although humanitarian operations may be viewed with mixed feelings organizationally, the military is excellent at dutifully executing national direction. If that direction is clearly to support humanitarian operations, the response can be delivered effectively with a single-minded purpose.

A common perception among NGOs is that "the military is not a humanitarian actor." They believe that when military forces provide assistance to a civilian population during a conflict, it is often to further the policies of their national governments, provide force protection, and meet their international legal obligations. Yet under exceptional circumstances, such as when civilian aid providers are unable to operate for security reasons, many civilian leaders acknowledge that military assistance is appropriate. However, many NGOs still weigh the risks associated with cooperating with military elements against what appears to be meager benefits and often opt to go it alone. They deem it to be more important to retain their image of independence and impartiality by avoiding association with the military.

27

There are, however, two key roles that humanitarian organizations often want the military to play,[10] whether it is a UN peacekeeping force, a national armed force or, in the absence of both, local militia:

- <u>Maintenance of a secure environment in which aid can be delivered on a neutral and impartial basis.</u> What this means in practice needs to be developed on the ground. For instance, it does not normally mean providing armed convoys, because the need for armed convoys demonstrates that the environment is not secure.

- <u>Protection of civilians as required by international humanitarian law and, where applicable, human rights law.</u> This is increasingly a concern for many NGOs, particularly those who adopt a rights-based approach. It requires the military to take a more active approach to ensuring the rights of communities where they are deployed.

Where appropriate, the UN and NGOs may request military support, particularly from a peacekeeping force. This may be necessary for logistical reasons: transport of personnel to inaccessible areas, transport of material supplies for humanitarian relief, or rehabilitation of roads, bridges, and culverts.

When the operational focus shifts from humanitarian assistance to reconstruction, civilian and military roles may become blurred, and interference with each other's efforts can be a problem. The military often lacks a long-term development focus and will sacrifice sustainability for speedy results. Moreover, the military generally has little or no training or expertise in international development. It is often criticized for conducting redundant assessments and inadequately coordinating with civilian elements, leading to duplication of effort. The military also may discount the long-term capacity of the local population to sustain development projects. Civilian groups perceive that military reconstruction projects would be more effective if they fully leveraged the military's unique engineering capabilities to repair basic infrastructure such as roads, bridges, power stations, and water supply systems. This would allow civilian agencies to do their own work more effectively and contribute to a sense of good will among the local populace by facilitating the restart of commercial activity and employment.

Establishing Humanitarian Space

Having identified differences between military and civilian operating cultures is not the same as saying cooperation and sharing of information are impossible or inadvisable. Military objectives are driven by political objectives, and humanitarian actions are driven by concern for the affected population. Where these objectives and concerns coincide, effective civil-military cooperation can be established. In fact, it is neither uncommon nor unlikely for militaries to be given the task of carrying out a political objective, such as the creation of a stable political environment with full respect for human rights, which dovetails with humanitarian objectives.

---

[10] Currion.

28

When the affected national population is a military target, however, cooperation is virtually impossible for humanitarian actors. As a result, a key factor in the deployment of civilian organizations is establishing and maintaining a conducive humanitarian operating environment. This is sometimes referred to as "humanitarian space" and is defined as the independence, flexibility, and freedom of action necessary to gain access and provide assistance to beneficiaries in a humanitarian emergency.

To effectively promote the respective missions of the military and the civilian organizations, each must have their own spheres of operation or "spaces" and information integrity. They need to coordinate work toward the common goal of protecting and helping the local population and leaders when their needs and objectives overlap. The civilian and military elements must share suitable, protected mechanisms for exchanging information. Effective coordination is essential to establish and protect the "humanitarian space" and to meet the collective capacity-building objectives of the affected nation.

The Civilian Information Environment

Humanity, Neutrality, Impartiality

To achieve the goal of alleviating suffering wherever it is found, all parties must perceive that aid workers and organizations are adhering to the key operating principles of humanity, neutrality, and impartiality. Consequently, maintaining a clear distinction between the roles of humanitarian actors and military actors is often the determining factor in creating an operating environment in which humnitarian organizations can discharge their responsibilities effectively and safely. Sustained humanitarian access to the affected population is ensured when the receipt of humanitarian assistance is not conditioned on any profession of allegiance to a party in a conflict. Humanitarian assistance should be perceived as a right independent of military and political allegiance.

UN General Assembly Resolution 46/182 offers the following specific definitions of humanity, neutrality, and impartiality:

> Humanity: Human suffering must be addressed wherever it is found, with particular attention to the most vulnerable in the population, such as children, women, and the elderly. The dignity and rights of all victims must be respected and protected.

> Neutrality: Humanitarian assistance must be provided without engaging in hostilities or taking sides in controversies of a political, religious, or ideological nature.

> Impartiality: Humanitarian assistance must be provided without discriminating as to ethnic origin, gender, nationality, political opinions, race, or religion. Relief of suffering must be guided solely by needs, and priority must be given to the most urgent cases of distress.

In addition to these three humanitarian principles, the UN seeks to provide humanitarian assistance with full respect for the sovereignty of states. As also stated in General Assembly Resolution 46/182:

> The sovereignty, territorial integrity and national unity of States must be fully respected in accordance with the Charter of the United Nations. In this context, humanitarian assistance should be provided with the consent of the affected country and in principle on the basis of an appeal by the affected country.

Information Sharing among Civilian Groups

Information sharing is often less restrictive among civilian aid workers and organizations than between civilian and military groups. Civilian groups use Internet web sites and portals extensively to share information. But their implementation of ICT capabilities is usually *ad hoc,* and information management and control are less structured than in the military. NGOs will frequently form their own formal or informal coordination bodies and monitor each other's activities, since bad practices by one NGO make the entire NGO community look bad. Members of an NGO community frequently decide on joint positions on key issues, and then use peer pressure to ensure that the entire NGO community adopts those positions.

The most active coordination mechanisms are normally sectoral (functional area) groups that focus on a particular topic, such as refugee camp management. In these groups, very strong coordination is frequently in place. It is at the broader, policy level that coordination tends to be weaker, although the UN normally takes the lead at this level. It is worth remembering that large variations can occur between different NGOs. Some NGOs are large, professional organizations with a long history in humanitarian work. Some are very small and not as well organized. NGOs have no obligation to coordinate with each other, and the different character of each NGO gives it a different approach to coordination.

Meanwhile, there continues to be a reluctance to share information with the military, due to the civilian groups' desire to remain independent and neutral. Private aid groups do not want to be perceived by the population and affected nation's leadership as an intelligence-gathering arm of the military. Additionally, when information is provided to the military, it is frequently absorbed into the classified system, making it unavailable to civilians—even to the group that originally provided it.

Yet, it is clear that HADR and S&R operations could benefit immensely if both military and civilian responders contributed and shared information needed in the humanitarian space. What is required is a protocol for providing information that does not compromise the goals of the overall humanitarian effort.

The Military Information Environment

In modern military forces, communications and information management are components of a broader concept referred to by some forces as *command, control, communications, computers, and intelligence* (C4I). Military communications and information technology (IT) systems are designed to ensure that the chain of command, essential control functions, and the intelligence process extend throughout the military force. These dedicated internal systems are among the most vital in any military operation. If an adversary can disrupt, damage, or destroy these systems, the ability of the force to function and even survive is at risk.

To protect the integrity of these systems, access to military communications and information management systems is carefully controlled. Levels of access to both communications channels and the actual information in the system are limited. Access to information is managed through a system of "classification" that determines whether information is sensitive or not. All military personnel have security clearances that determine the level of information to which they have access. Within a level, their access is further restricted by their "need to know." In other words, they are only given access to the sensitive information they need to know to perform their jobs.

Military personnel are consistently reminded to maintain "information security" and "operational security." The latter refers to protection of the intentions, plans, and capabilities of the forces. Thus, information is linked to the security of the force and denying potential adversaries knowledge about the force. As a result, military personnel, as a matter of policy and training, are hesitant to share information. Access to facilities where this information is collated, shared, or disseminated is strictly controlled.

Military commanders, even at the highest level, have limited authority to share classified information with personnel who have not been vetted. Information received from intelligence agencies is often classified and controlled by the intelligence agencies. Commanders in the field normally have no authority to share this information beyond approved addressees, even within their own organizations.

Among the most important of controlled facilities is the "operations center" in a headquarters or the "command post" at the lower tactical level. These locations are critical nodes in the C4I system. Access to these locations is always restricted and, in combat operations, military forces will attempt to keep these locations secret. Unescorted access to these areas is rare, even for personnel with the appropriate level of security clearance.

In situations where the military recognizes the need to share unclassified (non-sensitive) information with humanitarian and other civilian actors, this information will normally be shared via a *civil-military operations center* (CMOC) or, in NATO parlance, a *civil-military cooperation center* (CIMIC). Other information-sharing avenues may include the dispatching of military liaison officers to meetings, electronic bulletin boards, Internet web sites/portals, or even simple exchanges of e-mails.

Complex Operational Environments

The decision to intervene in a conflict is political and the military mission in support of the intervention reflects the political process. In a post-conflict or disaster response mode, however, the primary mission of the military is to create a safe and secure environment so that civilian government agencies, IOs, and NGOs, can conduct humanitarian assistance and assume appropriate responsibilities for civil policing, justice, governance, economic reconstruction, and nation-building activities. The military is not there to do the jobs of the civilian agencies and organizations.

However, when security is not established, the military may find itself obligated both to impose security and, temporarily, engage in crucial humanitarian assistance, governance, restoration of essential services, and other reconstruction assistance until the security environment allows civilians to take over these tasks. The civil-military mission is, in this circumstance, to enable the host country's leadership to establish the necessary capacity to manage governance, rule of law, reconstruction, and economic recovery.

These types of scenarios can be termed *complex environments*. In these environments, information expectations between the military and the civilian elements must be carefully managed. The military has been repeatedly told that the humanitarians have superior knowledge of the humanitarian situation, culture, language, and the population in general. Humanitarians have grown to believe that the vast intelligence capabilities of modern states and militaries are available to all military units. Inordinate mutual expectations can lead to the belief that information is intentionally being withheld and that erroneous information has been intended as disinformation. Achieving a shared civil-military vision, managing shared expectations, and facilitating collaboration, coordination, and information sharing is crucial to achieving unity of effort in complex emergencies.

Some progress has been made in establishing multinational collaborative information environments for coalition military operations. For example, the U.S. Government's secure multinational military information network—the Coalition Enterprise Regional Information Exchange System (CENTRIXS) has been employed to create multilevel secure virtual private networks (VPNs). Although the U.S. military controls the release of information and network access privileges to this network, it does create a facility for collaboration and information sharing among the U.S. military and other militaries. Currently, no similar formal arrangement is in place for sharing information among civil-military elements responding to a complex emergency, HADR, or S&R situation. The Internet does offer a medium for such sharing, and it has been used to create some limited collaborative arrangements to share civil-military information.

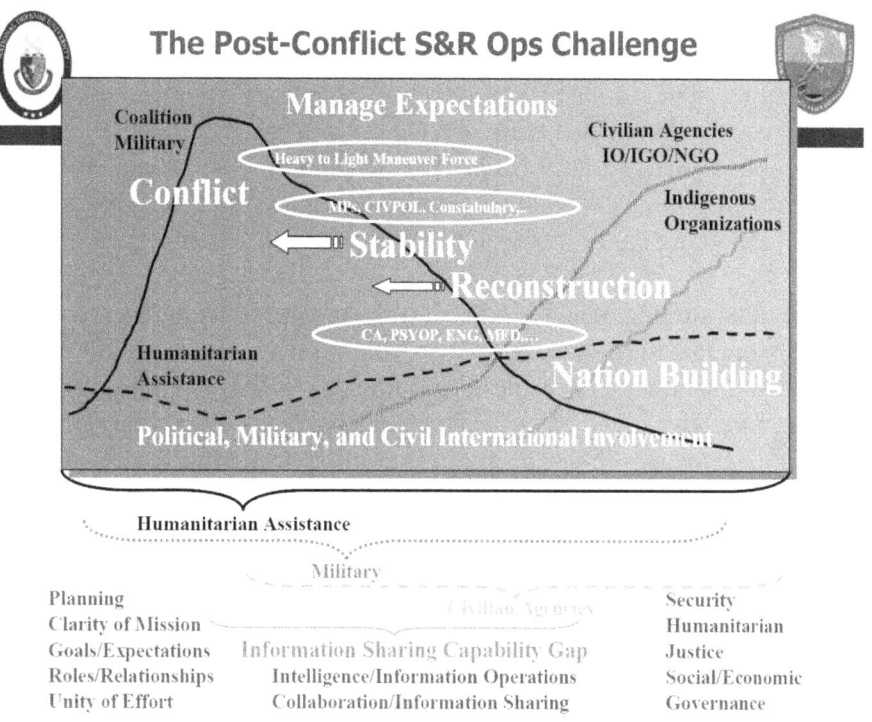

Figure 5

Building toward Civil-Military Coordination

During the transition from combat to S&R operations, it is necessary to adjust military force structure and capabilities. In today's high-tech operational environment, however, demarcation of distinct phases is highly unlikely. Figure 5 highlights the blurring of the phases, with lingering combat operations still being conducted, even as civilian and military S&R operations proceed—as in Afghanistan and Iraq. A heavy combat force may be redeployed or configured to a smaller forcer, better able to operate in urban areas. Force augmentation or adjustments need to address coexisting civilian security, counter-insurgency, counter-terrorism, and organized crime security and law enforcement needs. Additionally, S&R activities must be addressed, including: humanitarian assistance, such as food, clothing, and shelter for internally displaced persons (IDPs) and refugees; restoration of emergency services, such as fire and rescue, hospital, ambulance; infrastructure repair (power, water, transportation, communications, sewage); governance; health care services; education; and employment.

If one considers the entire operational spectrum depicted in figure 5, both military and civilian resources are employed across the full spectrum. Both the military and civilian agencies have their own core competencies. Trying to sub-divide the spectrum accomplishes little and instead reinforces the view that easily discernable and distinct thresholds separate civilian and military operations.

It should also be recognized that the civilians are likely to be present before the military is called in and will remain long after the military has departed. A number of transitions will occur over time, and ultimately, the intent is to hand off complete and self-sustaining peace to the affected nation. What really occurs in today's operational environment is a coalition of coalitions that are active in the affected nation. Civilian authorities operate where they can while combat is ongoing, and when it terminates, they assume responsibility for helping build the capacity of the affected nation.

Often, these civilian-military coalitions are formed in peacetime, or before a disaster strikes. Security organizations like, NATO or ASEAN may work together with development organizations, such as the World Bank or WHO, as part of "standard operating procedure" in a given country. Then, when a crisis or disaster occurs, these coalitions shift their focus—either toward post-conflict S&R (if armed conflict has occurred) or post-disaster humanitarian assistance and relief work. Not all crises necessarily require military force to resolve. For example, the very real economic crisis in the Pacific Rim countries during the late 1990s was resolved in most countries by civilian agencies.

The key point is that these civil and military coalitions will vary as tasks change and resources are adjusted to accommodate circumstances. All partners should be involved with the planning and execution of operations throughout the full spectrum of activities. To try and separate these interactions and ignore the need for coordination seems counterproductive. Moreover, the desired end state is a "political-military" solution, not merely a military solution.

ICT Support Challenges

Connectivity into a crisis area has two aspects: the technical capacity to build the ICT infrastructure and the organizational and inter-organizational structures to make sure the ICT capability is used effectively and appropriately. Both rely on establishing standards and policies that create a common platform for connectivity. Additionally, a collective agreement on the approach to be taken should be made prior to deployment, not during the deployment, as is often the case.

Such agreements need to cover a range of issues, including the question of who owns and controls the ICT infrastructure. This is particularly important for relations with affected host-nation governments. For example, national military and civilian agencies and UN organizations deploy by invitation of the host government, or under an international mandate that allows them to set up and operate their ICT capabilities. But regulations and requirements in the affected nation can make it extremely difficult for NGOs or other civilian organizations not operating under such a mandate to import, rapidly deploy, and use their ICT equipment without prior consent of the local authorities.

Civilian and military communications and information systems supporting emergency or S&R operations tend to have limited coverage and capacity. Moreover, networks commonly are not interconnected with each other. So the commercial cellular and satellite phone networks and Internet usually become the *de facto* communications and information networks linking civilian and military units. The potential effects of relying on these networks (where they are available) should be closely monitored, however, to make sure that over-reliance on any one network does not jeopardize the overall effort. For example, use of the Internet exclusively may exclude from coordination all parties who cannot receive email. Also, the surge of disaster responders using local cellular networks can collapse the capacity of those networks, to the detriment of all. Although commercial ICT services offer a means to create a civil-military collaborative information environment (CIE), policymakers on both sides of the civilian-military divide need to evaluate the necessities and realities they face in the field and work toward ways to balance the need for information security with the need to collaborate, coordinate, and share information.

Civil-military coordination challenges span differences in culture, language, organization, training and education, doctrine, planning and analysis, and communications and information systems. Unfortunately, old business models and restrictive policies continue to be applied to support what have become highly dynamic, collaborative needs and requirements in complex operations. Informal and unofficial personal relationships usually play an important role in cutting through this organizational "noise" to achieve information sharing. But the informal nature of this process makes it difficult to serve a wider audience or to institutionalize any standard procedures. As a result, "sneaker nets" (that is, hand-carried messages) and face-to-face meetings often become the only effective means to collaborate, coordinate, and share information in complex environments and post-disaster situations.

In spite of multiple "lessons learned" reports from past experiences, many of the same issues continue to plague the civil-military response community in each deployment:

- Little or no "shared informational awareness" to enable a common understanding of what data and information are available, what has been or needs to be done to it, who needs it, or even who has it;
- Multiple organizations producing the same information products;
- Organizational use of deprecated (i.e. old) data;
- Stove-piped and incompatible ICT systems that do not allow sharing of information with others, either because of format incompatibility or bandwidth and connectivity inadequacies;
- Large amounts of data being pushed to multiple locations, at multiple times, clogging up network bandwidth;
- Inadequate bandwidth access for many end users, who cannot support downloading large data files such as maps and imaging files; and
- Lack of tools, indexes, and metadata markers that would allow those in the field to find needed information or even be aware of what information might be available to them.

Effectively collecting, compiling, analyzing, and disseminating timely and relevant information is one of the primary challenges for S&R and HADR response operations.

Better information and knowledge management can improve the effectiveness of assistance. The faster civil-military response organizations can identify, collect, analyze, and disseminate critical information, the more effective the response becomes.

When HADR, S&R, or complex emergency deployments call for a military element, the military responds by implementing both classified and unclassified networks to support its C4I, logistics and other information exchange needs. Military access to the Internet is provided and websites related to the emergency appear on both the classified and unclassified networks.

For their part, the civilian elements are increasingly more capable of bringing with them the necessary ICTs to support their mission needs. The larger civilian organizations can deploy significant commercial ICT capabilities to crisis response areas. For example, USAID Disaster Assistance Response Team (DART) teams employ flyaway ICT capability packages and establish in-country Very Small Aperture Terminal (VSAT) networks to satisfy both initial-response and longer-term needs.

UN agencies such as UNHCR, UNICEF, WFP, and UNDP manage their own fixed global information networks and can employ either their own flyaway ICT packages or, on a case-by-case basis, can employ the WFP Dubai-based Fast IT and Telecommunications Emergency and Support Team (FITTEST) to extend and manage secure ICT services for UN operations in crisis areas. UNDPKO manages its own secure global ICT network and suite of flyaway packages tailored to support the needs of peacekeeping operations on the ground (and for coordination with UN headquarters). The civilian government and UN agencies also create websites to share information, maps, and Geographic Information System (GIS) products. Although other civilian organizations may deploy with lesser ICT capabilities, they, too, can often access the Internet and create websites to share information.

Figure 6 provides an example of how much of the information flow is now unofficial and decentralized. Bloggers populate the Internet with information on each humanitarian crisis. Additionally, "virtual" information and operations centers are emerging through the use of multifunction portals that provide metadata registries and catalogs, links to relevant web sites, and shared workspaces, providing "one stop shopping" for information and data.

Civil-Military S&R ICT Responses

Internet is the "de facto" civil-military collaborative information network

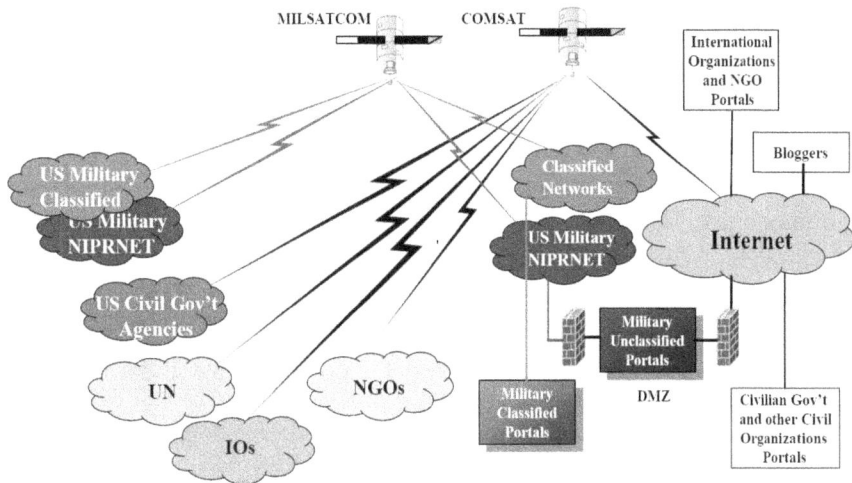

Figure 6

The proliferation of complex operations-related information websites and portals has helped policymakers at higher levels of government and civilian organizations. But it has become a major challenge—and even impediment—for responders on the ground. In many instances, the plethora of portals merely complicates the task of finding and obtaining useful information. Many websites seem more oriented towards informing the broader population by gathering, assessing, and summarizing information from many sources rather than analyzing and packaging information to more effectively respond to needs in the field.

Populating Internet web sites with data is useful in the field only if users can quickly find the data, understand its content and value to their operation, and then download it—usually over a low-quality, narrow-bandwidth communications link. Responders on the ground have neither the staff nor the time to spend hours surfing the Web. What they need is to receive specific information, quickly, from a trusted source.

Field workers also have problems when data are not adequately packaged, summarized, and tagged to allow quick assessment of their utility. Better tagging of data and better tools are needed to enable more effective and targeted searching for useful information. Guidelines in developing the content stored on websites, including the geo-referencing of data, where appropriate, are also required.

Simple approaches would also fill a void, such as an analysis cell that evaluates, packages, and tags information of use to the responders. Additionally, the creation of a "one-stop shopping" portal to provide a trusted information source would be helpful. Such a portal would store data and information, provide links to other relevant websites, offer shared working spaces and chat rooms, and give field workers the ability to easily find relevant information and data.

Commercial ICT Support Solutions

Commercial ICT capabilities are becoming more pervasive with each new S&R and HADR operation. Civilian and military elements deploy with laptops, cell phones, and, often, satellite phones. They use "personal digital assistants" (PDAs), handheld line-of-sight radios, and HF, VHF, and UHF radios to communicate. Ground-to-air radios are used to coordinate helicopter operations. Land mobile radios are used for emergency service communications. To the extent possible, participants also use the public switched telephone network (PSTN) and cellular networks to communicate and access the Internet through local Internet service providers (ISPs)—if they exist.

These communications systems, however, are often limited by a lack of surge capacity and inadequate coverage—particularly after infrastructure has been damaged in conflict or a natural disaster. Modular, packaged systems are often flown in to replace the damaged or destroyed commercial networks. In some cases, commercial "cellular on wheels" (COWs) networks, transportable transmission systems and WiFi "hot spots" are deployed, providing additional capacity and coverage. Commercial satellite systems are used for voice communications and remote access to the Internet. Systems such as Inmarsat, Iridium, Thuraya, and Globalstar provide voice and limited data service from small, often handheld, terminals. For larger data flows, VSAT links and networks are set up to support temporary fixed-installation needs, such as information and coordination centers. Inmarsat Mini-M, BGAN, and RBGAN satellite terminals are used to establish remote satellite access to telephone and Internet service.

Global Positioning System (GPS) receivers are used not only for location identification and navigation but to geo-reference data collected in the field. GIS and mapping tools are used for assessments and visualization of situation awareness.

Collaboration software tools such as NetMeeting, Groove, SharePoint, and InformationWorkSpace are used to create shared workspaces online, linked by wireless or wired LANs, which are then linked to the Internet via commercial satellite access. Along with other tools and Internet access, collaboration tools can help enable responders to make the most productive use of email, Web surfing, limited data exchange, and video conferencing. VoIP applications such as Skype are used for voice communication via the Internet and other Internet Protocol (IP) networks.

Several NGOs and businesses now specialize in providing information and telecommunications packages for field deployments. These packages can encompass ICT capabilities ranging from deployable commercial satellite communications and information capability packages to "Internet-in-a-box" or turn-key information services that are compatible with host-nation ICT projects such as e-government services.

The real challenge is not finding the components of an ICT deployment in HADR, S&R, and complex emergency scenarios, but rather in finding a rational management solution to linking, coordinating, and making the best use of the systems that are set up by multiple responders, civilian and military. No one organization is responsible for all IT configuration and network management, information management, information assurance, and system controls. *Ad hoc* methods—often little more than a stack of scribbled notes held together by a paper clip—are used to create telephone directories, email address listings, and lists of what systems are being deployed in what locations. Without any control or coordination, problems can arise, such as when two WiFi networks are deployed on the same frequencies, requiring de-confliction.

No mutually agreed, global concept of operations (CONOPS) or system architecture for ICT support to S&R or HADR operations is available. Furthermore, no person or entity in the civil-military community has stepped up and claimed responsibility for pulling together the disparate capabilities and creating and managing a federated ICT network and a distributed information and knowledge environment. Every deployment is largely an *ad hoc* event that employs "old boy" networks and personal contacts to create a workable mode of operations.

The result is a loss of efficiency that often threatens to become a loss of effectiveness—a disturbing prospect in the context of operations that directly lead either to saving or losing lives. There would thus seem to be real imperatives to set standards for collaboration, coordination, and information sharing and to identify the technological means to achieve them using ICTs.

Guiding Principles

Perhaps the place to start attacking the issue of contingency ICT support is to identify, based on previous experiences, several basic principles regarding planning and executing complex, multinational civil-military operations in man-made and natural crises. In general:

- Currently, disconnected operations are the rule, rather than the exception.
- Agility is more valuable than planning, but planning is essential.
- Effectiveness depends on fully incorporating non-military organizations.
- Trust grows better when collaboration occurs on neutral ground.
- On neutral ground, commercial off-the-shelf (COTS) solutions protect the data—not the network.
- Vague and overlapping boundaries exist between war and peace.
- Humanitarian organizations' support must begin as soon as the smoke clears.
- In civil-military information sharing:
- culture and social change are challenges,
  - policy and funding arrangements need updating, and
  - technology is an enabler.
- The Internet is the de facto medium for a civil-military CIE.
- Civilian and military ICT capabilities share commonalities stemming from deployment of commercial ICT products.
- The large number of new commercial products constantly emerging offers opportunities to make smarter use of deployed communications and IT capabilities.
- Perhaps the greatest challenge is creating a culture of information sharing that:
  - promotes the free flow of data, information, and ideas;
  - facilitates informed decisionmaking; and
  - builds trust and commitment among stakeholders.
- Information management is not a set of discrete tasks but rather a process that:
  - requires strong leadership, vision, and investment;
  - underpins all aspects of civil-military response;
  - requires long-term institutional support and ample, sustained investment.
- Technology needs to be adapted to the changing dynamics within and between the groups that use the technology.
- Beyond these general principles, more specific principles can be sketched out, addressing particular aspects of ICT support.

USAID's Nine Principles of Reconstruction and Development[11]

The U.S. Government has made development work a national security priority; the September 2002 *National Security Strategy* underscores development as one of three

---

[11] Andrew S. Natsios, "The Nine Principles of Reconstruction and Development," *Parameters*, (Autumn 2005), vol XXXV, no. 3.

strategic areas of emphasis (along with diplomacy and defense), and clearly states that "including all of the world's poor in an expanding circle of development—and opportunity—is a moral imperative and one of the top priorities of U.S. international policy."[12]

This new development climate has led USAID to determine that it requires a more uniform and consistent set of guiding principles, and that these principles must accurately reflect how USAID approaches development from all levels—from day-to-day project operations to high-level policy decisions. USAID has completed a series of recent policy strategy documents, including "U.S. Foreign Aid: Meeting the Challenges of the Twenty-first Century and the Fragile States Strategy."[13] Based partly on these documents, USAID has developed the following "Nine Principles of Reconstruction and Development" to guide its efforts:

Principle 1: Ownership. Build on the leadership, participation, and commitment of a country and its people.
Principle 2: Capacity Building. Strengthen local institutions, transfer technical skills, and promote appropriate policies.
Principle 3: Sustainability. Design programs to ensure that their impact endures.
Principle 4: Selectivity Allocate resources based on need, local commitment, and foreign policy interests.
Principle 5: Assessment. Conduct careful research, adapt best practices, and design for local conditions.
Principle 6: Results. Direct resources to achieve clearly defined, measurable, and strategically focused objectives.
Principle 7: Partnership. Collaborate closely with governments, communities, donors, non-profit organizations, the private sector, IOs, and universities.
Principle 8: Flexibility. Adjust to changing conditions, take advantage of opportunities, and maximize efficiency.
Principle 9: Accountability. Design accountability and transparency into systems and build effective checks and balances to guard against corruption.

The Nine Principles are a formalization of customary USAID operating procedures. They reflect key institutional principles that most seasoned aid agencies incorporate in all their work, from ensuring local ownership and sustainability of a health clinic to flexibly adjusting a rural development program to counteract poppy cultivation. (One exception may be the selectivity principle, which for USAID takes into account U.S. foreign policy objectives; such objectives are not always built into aid agencies' priorities.)

The Nine Principles significantly overlap with military doctrinal principles. The continued development of the military S&R operations platform and the increasing

---

[12] *The National Security Strategy of the United States of America* (Washington: The White House, September 2002), 21, available at http://www.whitehouse.gov/nsc/nss.pdf.
[13] USAID, "White Paper: U.S. Foreign Aid, Meeting the Challenges of the Twenty-first Century," January 2004, http://www.usaid.gov/policy/pdabz3221.pdf (hereinafter, "White Paper"); USAID, "Fragile States Strategy," January 2005, http://www.usaid.gov/policy/2005_fragile_states_strategy.pdf.

frequency of civil-military collaborations mean this convergence is here to stay. Additionally, effective reconstruction and development work cannot afford to overlook the Nine Principles. Quite simply, reconstruction is not effective when the local population does not feel a sense of ownership toward donor programs. Likewise, if donors ignore the accountability principle, not only does this set a poor example for the local population, but the legitimacy of the donor's overall involvement is brought into question.

Principles for Developing an Information Strategy[14]

A strategy is not a physical document. It should be a set of guiding principles and practices that the various actors voluntarily choose to incorporate into their work. The following points should be kept in mind. An information strategy:

- Must be related to functional area strategies (this provides it with the framework for implementation, as information management involves financial, human, and IT resources);
- Must have clear and realistic objectives that relate very clearly to the needs that have been identified by functional area working groups;
- Requires the support of senior management;
- Should clearly identify different types of information and the different purposes that this information is used for (security updates, baseline data collection, and situation reports); and
- Should recognize that different types of information have different uses for different actors and seek to maximize multi-purpose information.

A strategy document can be a useful guide to the strategy, but it should not be mistaken for the strategy itself. Such a document might include the following text elements:

Introduction: This should explain the strategy as well as the rationale behind it and articulate the goals.
Background: a brief explanation of the context in which the strategy is being implemented, including an outline of the opportunities and problems addressed.
Information Needs: Identifies the key players' information requirements and how the strategy will meet them. There should be a discussion of what needs are included and which ones are not, within the scope of the strategy. This also should include an audit of existing information resources and a list of requirements based on that.
Roles and Responsibilities: Identifies the organizations and individuals directly responsible for creating, analyzing, disseminating, or using information. Their roles must be clearly described and accepted by the actors.
Implementation: A prioritization of implementation steps, identification of a team responsible for ensuring that those steps take place, and clear management plans for individual projects.

---

[14] Currion.

42

Review Process: The articulation of a process for monitoring and regularly reviewing the implementation of the strategy.

Information Management Principles[15] [16]

In the broadest sense, information management covers a wide range of issues, including data collection. Several core principles guide information management. The first principle involves *accessibility*. Humanitarian information and data should be made accessible to all humanitarian actors through easy-to-use formats and by translating information into common or local languages. Information and data for humanitarian purposes should be made widely available through a variety of online and offline distribution channels, including the media.

Another core principle of information management is *inclusiveness*. Multiple stakeholders—especially those representing the affected population—should base information management and exchange procedures on a system of collaboration, partnership, and sharing, with a high degree of participation and ownership. *Inter-operability* means that all sharable data and information should be made available through networks, systems, and formats that can be easily accessed and linked by humanitarian organizations.

Information must be more than just accessible. Users must be able to trust that information and base decisions on it. This calls for *accountability*. Users must be able to evaluate the reliability and credibility of data and information by knowing its source. Information providers should be responsible to their partners and stakeholders for the content they publish and disseminate. This goes hand-in-hand with *verifiability*. Information should be accurate, consistent, and based on sound methodologies, validated by external sources and analyzed within the proper contextual framework. Other information management principles include:

- Relevance: Information should be practical, flexible, responsive, and driven by operational and decisionmaking needs throughout all phases of a crisis.

- Objectivity: Information managers should consult a variety of sources when collecting and analyzing information so as to provide varied and balanced perspectives for addressing problems and recommending solutions.

- Humanity: Information should never be presented or used to distort, mislead, or cause harm to affected or at-risk populations, and information use should respect the dignity of victims.

---

[15] UN OCHA, Symposium on Best Practices in Humanitarian Information Exchange, Final Report, Palais des Nations, Geneva, Switzerland, 5-8 February 2002

[16] Currion.

43

- Timeliness: Information should be gathered, analyzed, and made available while it still matters; out-of-date information can be worse than useless—it can lead to misinformed decisions that endanger the lives of affected populations and relief workers.

- Sustainability: In HADR and S&R operations, information and data should be preserved, catalogued, and archived so that it can be retrieved for future use, such as for preparedness, analysis, lessons learned, and evaluation.

Civil-Military Coordination and Cooperation Principles

Coordination and cooperation amongst civil organizations and military units in HADR and S&R operations is, understandably, an immensely difficult and complex task. Practitioners must accept that any operation will involve countless logistical problems involving interoperability issues and redundancies. To mitigate such problems, however, planners must accept and understand the complexities involved and plan accordingly.

A first step in collaboration is identifying points of agreement and disagreement that can be compromised. Military and civil organizations often differ greatly in management style and culture, but common ground can be found. These points should drive the selection of objectives for the operation. In terms of implementation, agreed-upon objectives must coalesce into a common purpose among all civil and military groups involved.

Each group's mission should reinforce and complement the missions of the others. "Lessons learned" reports often warn about "re-inventing the wheel" in each new deployment or situation. Not only is it important to understand what procedures and tools have been created in the past, but also to avoid tasking two units with developing similar procedures and tools, or in other words, "inventing the wheel simultaneously." Planners should seek to eliminate such redundancies amongst groups, addressing in particular the strengths and capabilities of each.

Resource management is also critically important to successful operations. Since resources are always limited, planners must allocate funds and other assets efficiently, again, keeping complementary missions in mind and avoiding redundancy. Finally, planning for collaborations in HADR and S&R operations should look forward to disengagement. Complementary action should not foster inter-dependency among groups.

Principles for Creating a Deployable Capability

The civil-military ICT responses to a crisis will need to address capabilities to support communications and information exchange needs of the initial responders that arrive on the ground. If ICT capacity exists and is sufficiently functional and usable in the affected nation, it still may be necessary to install some surge capability to accommodate the increased demand for service created by the responders. It most likely will also be necessary to do some early recovery of ICT capacity to enable sector activities bolstering civil security (policing), governance, economic, and social well-being.

Early assessments of the state of the affected nation's ICT capacity will need to be conducted to determine what recovery and reconstruction actions need to be funded and implemented in the near-to-mid term. Longer term, ICT capacity-building will need investments to develop new capabilities such as e-government, e-education, and e-commerce. Even ICT capacity-building actions taken in the initial and middle phases, however, should be done with the objective to leave them behind for use as a means to jump-start the affected nation's ICT sector growth. Hence, these actions and investments should enable the evolution to a long-term vision for the host nation's communications and computing capabilities.

In addition, USAID, DOS, and DOD should champion a COTS "collaboration zone" architecture that accommodates the restrictions and requirements of major collaboration tools, including the lowest common denominator: web portal, Groove, IWS, Share Point, and others. A metadata repository (registry, catalog, shared space) needs to be created to provide a one-stop-shopping capability to facilitate collaboration, cooperation, assessment, packaging, and sharing of information. This registry should be located within a ".org" or ".gov.xx" network domain to maintain autonomy from military and U.S. government elements. This is a necessary foundation to build a trusted information zone that can be freely and openly used by both civilian and military entities.

The Tampere Convention[17]

Regulatory barriers that make it extremely difficult to import and rapidly deploy equipment without prior consent of the local authorities have often impeded the trans-border use of telecommunication equipment by humanitarian organizations. Victims of disasters soon should be able to benefit from faster and more effective rescue operations, thanks to the Tampere Convention on the Provision of Telecommunication Resources for Disaster Mitigation and Relief Operations. The treaty simplifies the use of life-saving telecommunication equipment.

The Tampere Convention calls on signatories to facilitate the provision of prompt telecommunication assistance to mitigate the impact of a disaster. It covers both the installation and operation of reliable, flexible telecommunication services. Regulatory barriers that impede the use of telecommunication resources for disasters are to be waived in disaster scenarios. These barriers include licensing requirements to use allocated radio frequencies, restrictions on the importation of telecommunication equipment, as well as limitations on the movement of humanitarian teams.

The Convention describes the procedures for requesting and providing telecommunication assistance, recognizing the rights of countries to direct, control, and coordinate assistance within their territories. It defines specific elements and aspects of

---

[17] International Telecommunication Union publication, "Telecommunications Saves Lives," available at: http://www.itu.int/ITU-D/emergencytelecoms/doc/brochure2.pdf.

telecommunication assistance, such as when that assistance should terminate. It requires governments to make an inventory of the resources—both human and material— available for disaster mitigation and relief, and to develop a telecommunication action plan that identifies the steps necessary to deploy those resources.

The Convention entered into force January 8, 2005. A complete status report is available by referring to the publication *Multilateral Treaties Deposited with the Secretary-General* in the *UN Treaty Collection.*[18]

---

[18] This report is on the Internet at http://untreaty.un.org/English/treaty.asp.

Part Two: TOOLKITS AND BEST PRACTICES

ICT Toolkits

As a result of ICT technology developments and expanding commercial offerings, a variety of flexible, scalable, and rapidly deployable civil-military ICT packages can and are being created, using off-the-shelf terrestrial wireless and commercial satellite products and services. Internet portals and metadata repositories also are being created to build distributed networks, to satisfy the needs of the civilian and military responders.

Telecommunications Networks

One of the major components of any ICT capability, of course, is the telecommunications network infrastructure that serves as the backbone and distribution system for voice and data transmissions. In disaster or post-conflict environments, demand for reliable telecommunications links does not decrease. To the contrary, relief and reconstruction efforts place additional demands on telecommunications capacity, at a critical time in the redevelopment of the sector.

Telecommunications Needs in S&R and HADR Operations

It is a reasonable assumption that the telecommunications infrastructure of the country in which an S&R or HADR operation takes place has been damaged. It may no longer have (if, indeed, it ever did have) sufficient capacity, bandwidth, or coverage to support the operational needs of all participants in a relief or reconstruction effort. Assisting militaries, civilian agencies, IGOs, IOs, and NGOs often bring their own capabilities to augment those of the remaining local telecommunications networks and to access commercial satellite communications services and the Internet.

Long haul, high-capacity systems using commercial satellite links (and fiber links where appropriate and available) will be employed to extend coverage and bolster data bandwidth. These will support the forward-deployed military C2, intelligence, and logistics requirements, as well as civilian humanitarian and reconstruction assistance needs. For example, USAID DART teams and the UN FITTEST flyaway ICT capability packages use commercial satellite links extensively to connect headquarters organizations with field elements. NGOs are now employing small ICT packages that rely on portable and mobile satellite terminals to access the Internet and global telecommunications services. Many of the civilian and military participants are using VSAT networks. Terrestrial radio systems (HF/VHF/UHF) and cellular phones are also used for field communications.

Sometimes, these capabilities can have unintended consequences when used by IOs and NGOs. In today's more hostile environment, Inmarsat phones and radio antennas on vehicles can make them a target for criminal and terror attacks. Reliance on cellular and satellite phones also can contribute to eroding the use of traditional secure radio networks that are still employed and used for emergency response communications.

Commercial ICT capabilities are becoming pervasive as part of both military and non-military inventories of deployable capabilities. Although commercial satellite capabilities provide wide-area, spot, and global access, coverage and capacity in any particular location can sometimes be limited, especially in remote areas. Responders should therefore not expect wideband services when they begin operations. Low bandwidth, intermittent coverage, and overloaded civil telecommunications and satellite infrastructures—which are not designed to absorb the surge in demand from responders—can significantly degrade end-user performance. These limitations need to be anticipated and planned for before deploying into a crisis area.

Satellite Communications

Perhaps the best place to begin discussing network architecture is with satellite connectivity—the field worker's window to the outside world. Commercial satellites today can provide cost-effective connectivity to virtually any place on Earth. As a result, they have become a key enabler for extending ICT services to remote, devastated, or disadvantaged areas in support of HADR and S&R operations. Portable, mobile, and fixed terminals communicate with land earth stations (referred to as "hubs" and "teleports") that interface with the public network for access to voice and Internet services, including email, Web surfing, and the use of Web portals.

Satellite phones and terminals include a range of options, from high mobility/low data-rate devices all the way up to fixed installations with higher bandwidth. The terminals include:

- Highly portable, low bit-rate (2.4 kilobits per second (kbps) to 9.6 kbps) handheld devices (satellite phones);
- Portable, low-to-medium bit-rate (2.4 kbps to 144 kbps) terminals from the Inmarsat family of satellite terminals also known as Mini-Ms; and
- Broadband (up to 8 megabits per second) mobile and fixed installations referred to as VSAT dishes.

Mobile satellite phones are similar in appearance and function to terrestrial cellular phones. They need direct, line-of sight access to the satellite, but because they use omni-directional antennas, they do not need to be aligned perfectly. Several satellite systems provide service for these types of phones:

- Thuraya – A single, geostationary satellite that provides limited coverage for about 100 countries (the coverage area includes Europe; North, Central, and parts of southern Africa; the Middle East; Central and South Asia, plus oceans in these regions);
- Iridium – A constellation of 66 low-earth-orbiting (LEO) satellites that provides global coverage;
- Globalstar – A constellation of 48 LEO satellites that provides nearly global coverage (but, due to the lack of gateway stations, does not work in Africa).

Moving up the bandwidth scale, a key provider of mobile, low-to-medium bit-rate global coverage is Inmarsat, a recently privatized consortium that operates four geostationary satellites covering the entire surface of the earth, except for the Polar Regions. Inmarsat, which began life as an IGO formed to facilitate maritime communications, provides a range of portable terminals, some of them "suitcase"-sized. Inmarsat's terminals must be operated in outdoor locations, within the line-of-sight of an Inmarsat satellite.

At the top end of the scale, several commercial satellite communications providers, such as Intelsat, PanAmSat, and New Skies, own and operate their own fleets of satellites and offer global broadband satellite coverage supporting VSAT connectivity. The medium-to-broadband VSAT antennas are installed in fixed outdoor locations, also within line-of-sight, and must be more accurately aligned to achieve high bandwidth connectivity. VSAT-based networks are typically arranged in a star-based topology, where each remote user is supported by a VSAT, which is linked to a hub or teleport that acts as the central node and employs a large-sized dish antenna with a high quality transceiver. The teleport is the gateway that provides connections to the public switched telephone network and to ISPs. VSAT networks are ideal for centralized, hub-and-spoke networks with a central host and a number of geographically dispersed terminals.

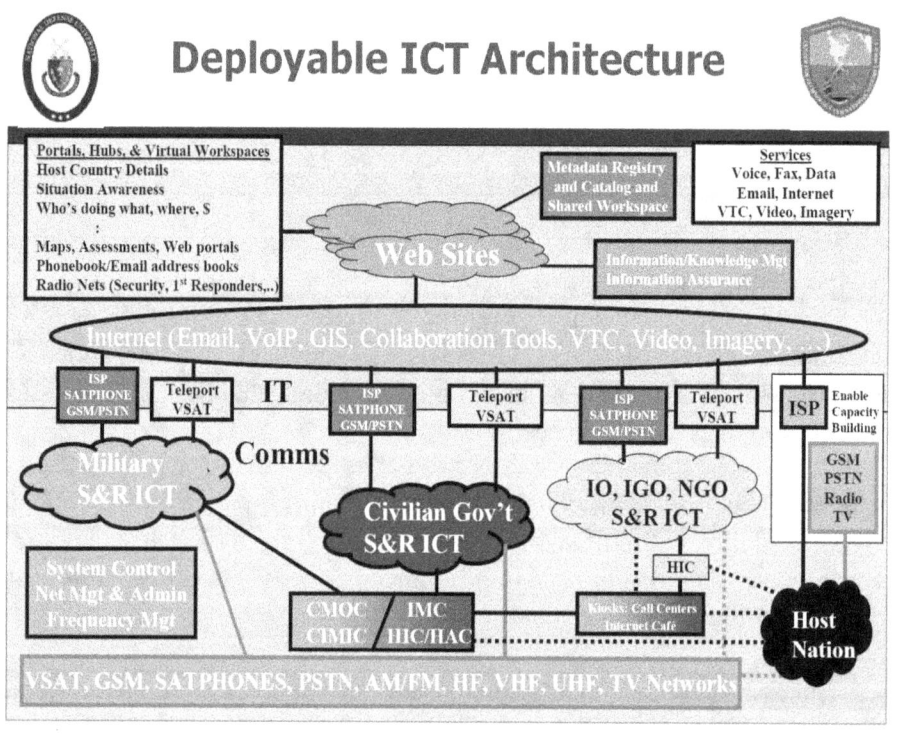

Figure 7

These satellite technologies can be used, in combination with terrestrial networks, to serve as the medium for civil-military coordination. Figure 7 illustrates what could be a "default" civil-military ICT architecture based on common practice today.

Creating A Collaborative Civil-Military Information Environment

Creating a common communications culture—increasing trust and setting a foundation for collaboration and information sharing—must be done without undermining either the neutrality of civilian IOs, IGOs, and NGOs or the need for the military to safeguard operational security information. A list that details who is doing what, where, is useful for resource allocation and management in relief and reconstruction. But in the wrong hands, it also can be a target list for groups and individuals that want to use violence for political ends. The creation of a *collaborative information environment* (CIE), such as that shown in figure 8, therefore, requires a balance between openness and security. It can be achieved only if everyone is sensitive to one another's concerns.

Perhaps the first step in building a CIE is to cultivate a willingness among the civil-military participants to create a common communications culture and to minimize barriers to participation, once it is established. Mechanisms to collaborate and share information can be created by:

- Using a common ICT response architecture (a federated network) employing the Internet, commercial ICT products and services, web portals, and metadata repositories;
- Creating a suite of interoperable ICT "toolkits";
- Extending ICT capabilities to a crisis area to be shared as appropriate;
- Facilitating and promoting collaboration and information sharing among participants;
- Supporting the building of host-nation ICT surge capacity and infrastructure recovery and reconstruction as needed;
- Forming civilian chief information officer (CIO) and military G6/C6 or equivalent collaboration arrangements, with needs assessment and requirements, planning and implementation, and federated network management and operation.

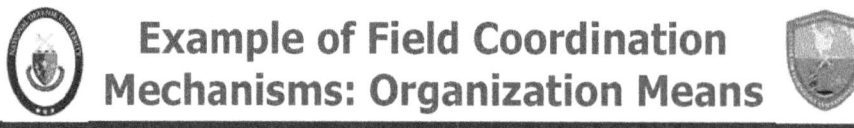

Example of Field Coordination Mechanisms: Organization Means

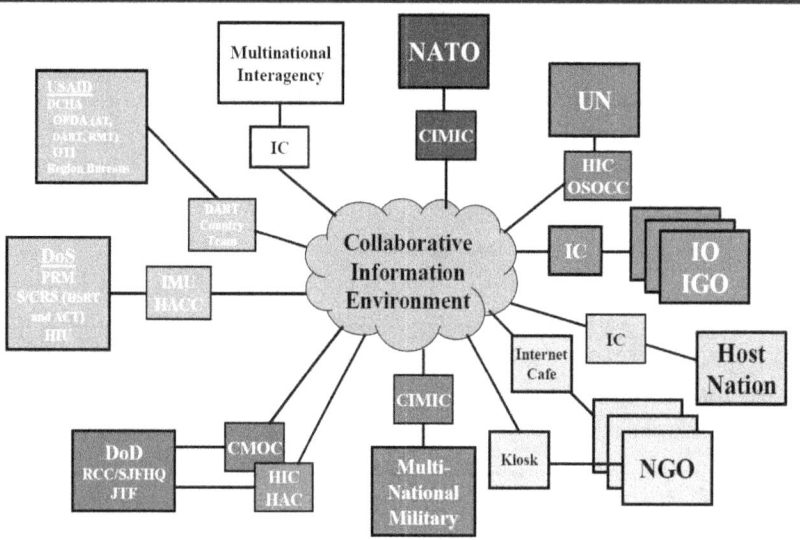

Figure 8

The ability to provide a ubiquitous ICT network, with the agility and adaptability for authorized participants to "plug and play" with their own equipment, means the CIE needs to respond effectively to event-driven, high-tempo, short time-scale, uncertain, diverse, and dynamically-varying operations. The CIE needs to be able to mediate flexible and timely interaction between "come-as-you-are" heterogeneous systems and information databases. It needs to support agile C2, shared situation awareness, and improved interoperability and collaboration.

The technical means used today to facilitate coordination within classified and unclassified information environments are illustrated in figure 9. The U.S. military classified and unclassified networks are largely based on employing secure guard gateways and strict information assurance processes to carefully manage and control access and use privileges. In contrast, the world's largest unclassified network, the Internet, operates in a highly trust-oriented operational environment with little, if any, information assurance measures.

Military and other agencies that produce classified products need mechanisms that allow them to more easily partition information, so that some of it can be routinely released to coalition military partners (at lower classification levels) and non-military parties, such as IOs and NGOs (fully declassified). There is also a need to more effectively deal with multiple levels of classification and the protection of dissemination within those multiple levels. The CIE, therefore, should be able to accommodate the management of access privileges, security, and performance. This has both collection and dissemination aspects in terms of access to, and use of, the CIE.

Field Coordination Mechanisms: Technical Means

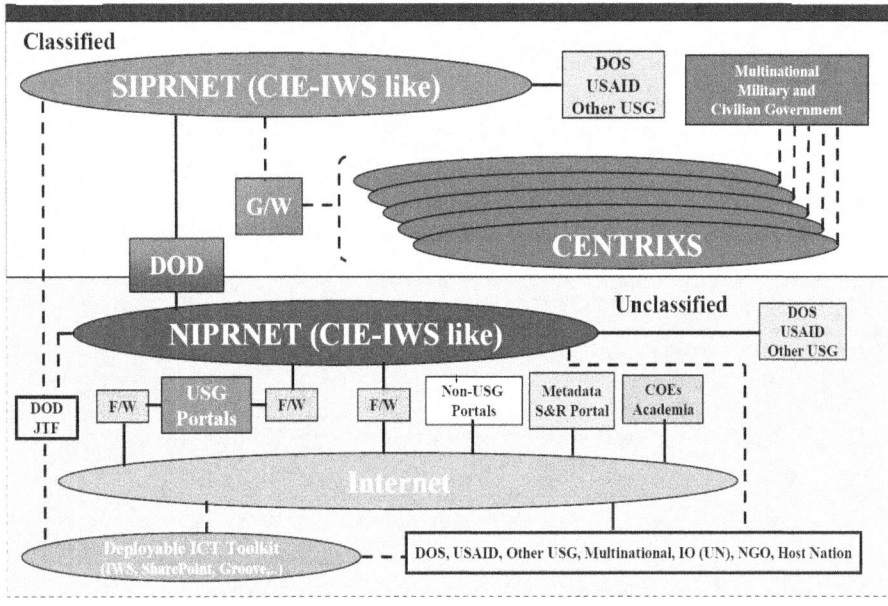

Figure 9

Language differences continue to challenge operational cooperation. Machine-translation tools are needed to help fill the translation gap. As noted, interoperability means not only technical and political compatibility, but also the will and the means to communicate, cooperate, and collaborate—in short, sharing a common culture of communication. When systems are not politically, organizationally, or technically interoperable, information becomes "stove-piped" within a single organization, and systems cannot easily collaborate.

The CIE needs to be able to accommodate a diversity of users, along with the diversity of organizational cultures, equipment, systems, and databases they bring with them. Data collection, decisionmaking tools, analysis capabilities, and visualization aids all must be improved to better meet the needs of S&R and HADR requirements. Included in this is the need for data and information management strategies that can be used to guide collection management, analysis, and database capabilities. Tools to support predictive assessment and course-of-action planning for S&R operations are needed as well, including measures of effectiveness (MOEs) and measures of performance (MOPs) for measuring progress and the success of actions.

The CIE needs to be scalable in terms of number and diversity of the civil and military users. Moreover, it must be easily and rapidly deployable into the area of operations and able to accommodate that environment. This may call for stand-alone capabilities including a power supply and operations and maintenance (O&M) support.

ICT Packages and Solutions

The creation of a collaborative military and civilian computing and communications environment is achievable with today's technology. An approach to creating a civil-military CIE is illustrated in figure 10.

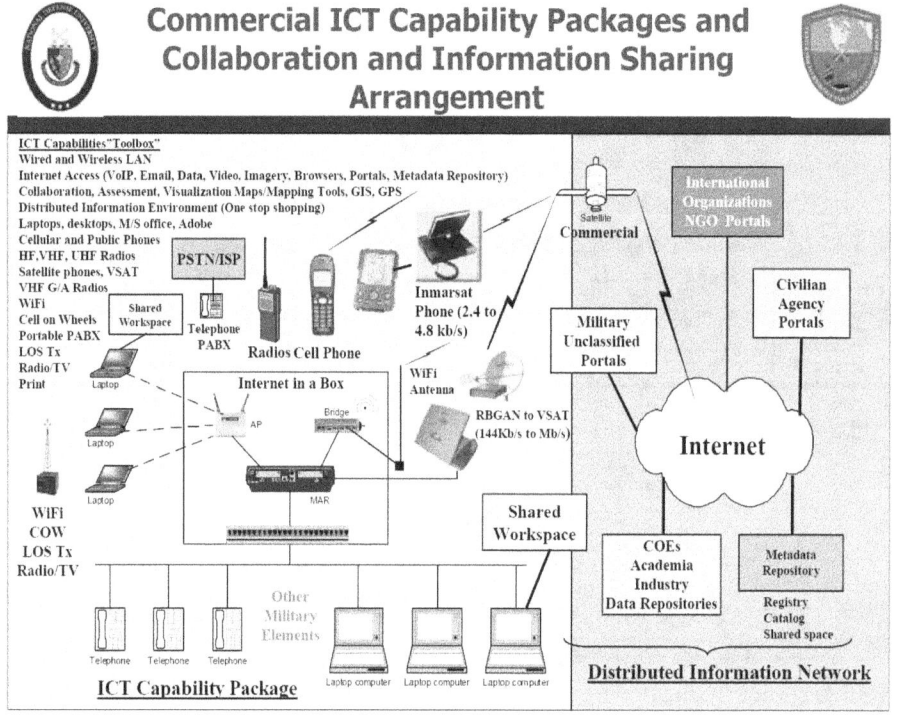

Figure 10

This approach is based on the following assumptions:

- The Internet, satellite links, and cellular phones will be the preferred media for communicating and sharing information among the civil-military participants.
- Commercial satellite service likely will be a primary means of gaining access from remote areas and to the Internet.
- A common suite of ICT capabilities, such as a "toolbox" containing cell phones, radios, satellite phones, VSATs, PDAs, laptops, workstations, WiFi networks, collaboration tools, GPS receivers, and GIS products, can be selectively packaged and tailored to meet the anticipated ICT operational support needs.

In some cases, the civilian and military responders will do the satellite access systems engineering, acquisition, and integration themselves. That is, they will determine what they need and then purchase or lease VSAT terminals, satellite phones, hubs, teleports and satellite access, as well as make the necessary Internet access arrangements and set up their own networks. Or, they will engage a satellite service provider and/or a systems integrator to do all this for them as illustrated in figure 11. Responders can take advantage of emerging NGOs and private companies that are now offering turnkey or managed satellite and Internet access services. Some packages may include the set-up and management of an information center in the operations area.

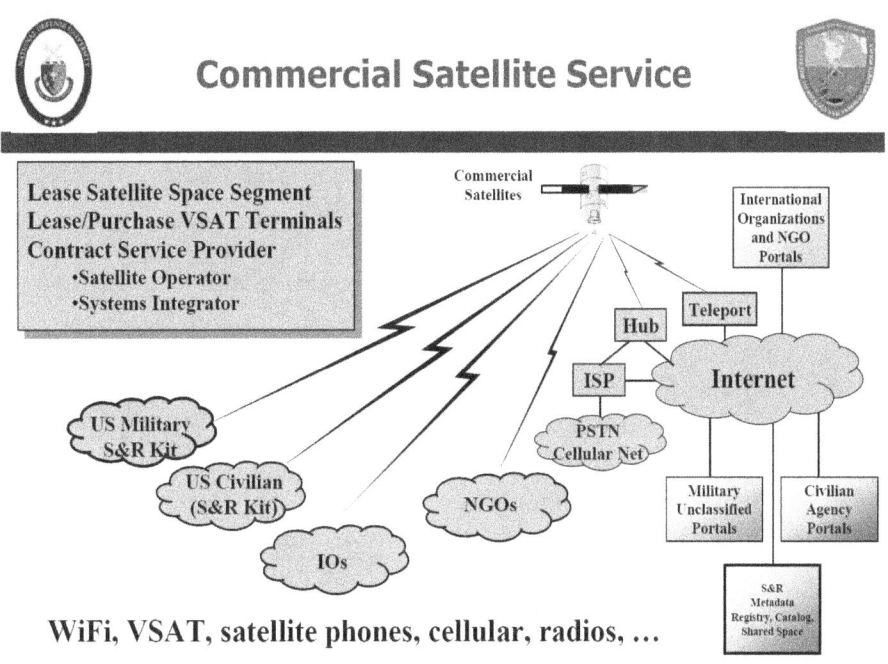

Figure 11

Adequate consideration must be given to local regulations governing the use of equipment and frequency assignments for wireless networks. Proper cellular phone, radio, WiFi, and satellite access configurations are vital. It is important to have local government approval for their usage, along with service agreements with satellite, telephone, Internet access, and teleport service providers before deploying ICT capabilities forward.

Common practice for many civilian and military entities today is to create Internet portals and metadata repositories in an effort to provide a "one-stop-shopping" capability for access to information about the crisis situation. Collaboration tool bridges will likely be needed to facilitate exchanges among different collaboration tool suites, such as InfoWorkSpace and Groove.

Presumably, the responder community will employ smart systems engineering to create appropriate local radio (HF, VHF, UHF) and cellular networks, facilitating communications among responders on the ground. Plus, agreements will be made for data standards and smart management of the information that will populate the various Web portals created to support the response operation, including the creation of a metadata repository to facilitate information discovery.

Information assurance will need to be more than "trust" and self-discipline by the user community. Appropriate security measures are required to protect and restrict unauthorized access to unclassified but operationally sensitive information. The community will need to have agreed mechanisms in place to insure the quality and integrity of information populating the web sites and to coordinate management of the resulting federated information network. Finally, appropriate network security and information assurance protection must be implemented to protect against intrusions, viruses, malicious code, and misuse of the network information.

Toward a Time-Phased Business Process Model

A jointly prepared civil-military ICT strategy should guide planning, not only for the initial deployment of ICT into a crisis area but also for the desired "end state"—the goals and objectives for ICT capacity-building and development—in any given nation. Early actions to build ICT surge capacity and reconstruct affected ICT infrastructures should include leave-behind technologies consistent with long-term capacity-building objectives. Figure 12 illustrates a notional business process model that could be used to guide the pre-planning, incident response planning (surge, recovery, and reconstruction), and planning for the "to be" ICT capacity of the affected nation.

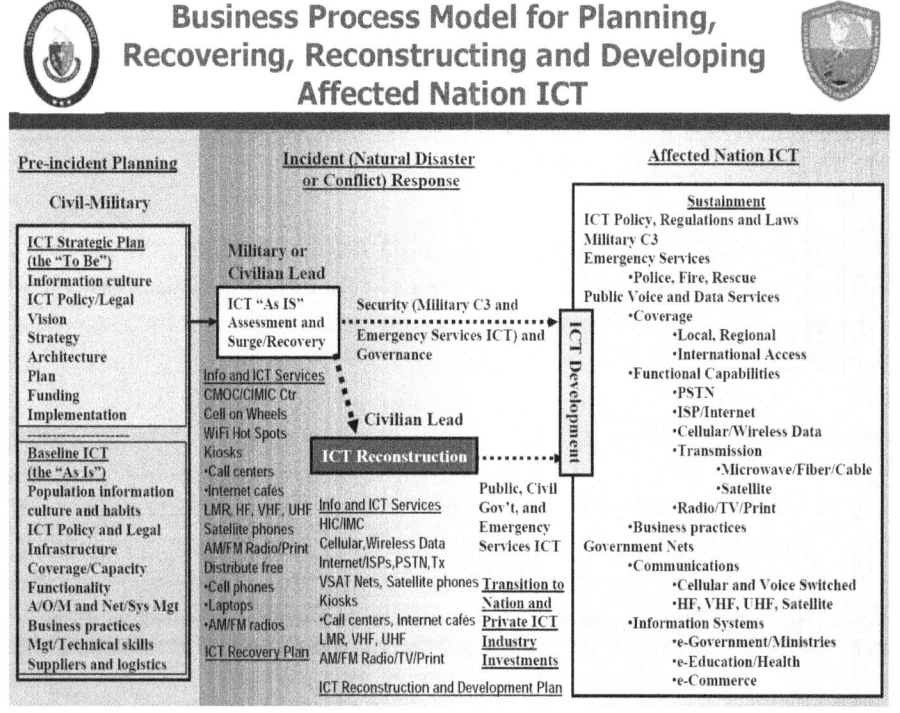

Figure 12

The pre-incident planning focuses on developing a baseline understanding of the ICT infrastructure capabilities and limitations of the host nation, including other ICT-related factors, such as existing regulations, policies, legal considerations, technical and management skills, economic considerations, and business practices. The vision for the "to be" ICT capabilities of the host nation needs to be based on the information culture of the nation and the government's views and desires for moving into an appropriate phase of the information age, whether that involves creating e-governance, e-commerce, e-education, e-economic, and/or other information capabilities.

Knowing the "as is" and the "to be," it is possible to develop an ICT strategic plan to guide investments, acquisitions, and other needed improvements, including regulation and legal changes and education and training needs. The incident response-planning phase needs to do an immediate quick look assessment of the ICT infrastructure to determine what works and does not following a natural or manmade disaster. This helps to determine what adjustments are needed to the ICT capability packages that will be deployed. It also can identify additional actions that may need to be taken to provide surge capacity and recovery of key ICT infrastructure, support the responder elements, and establish communications with the affected nation's leadership and population. In regard to the latter, kiosks with call centers and Internet cafes can be used, but it is also important to restore and use other means to conduct mass communications, such as radio, TV, and the print media.

Depending on the state of the affected nation's ICT infrastructure (pre-incident and post-incident), it may be necessary to develop both reconstruction and development plans to guide the ICT capacity building. If the infrastructure really never existed to start with or was totally destroyed, it may be more appropriate to go directly to planning for, and investing in, ICT development. Figure 13 illustrates a time-phased model of potential ICT development. The intent should be to enable the transition of the ICT capacity-building as soon as possible to the affected nation and, as appropriate and desired by the affected nation, to private industry.

Time Phased Situation

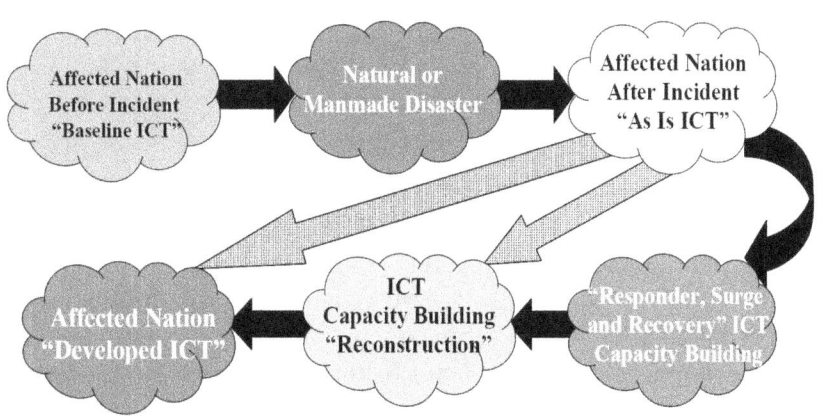

Figure 13

Short-Term Surge and Recovery

In the short- to mid-term, several different options are possible for deploying ICT capabilities and establishing CIEs. Responder organizations can establish ICT networks and services to support crisis area operations using:

- Commercial satellite service from satellite operators and systems integrators;
- Leased connectivity, services, and capabilities;
- Acquired ICT response kits from donors and/or NGOs;
- Kits assembled from purchased commercial products and access services;
- Self-provided networks; or
- A combination of the above.

Military and civilian elements can use Internet portals and metadata repositories to create a distributed information environment and, in the process of doing so, should use open system standards to the extent possible. In austere situations, amateur radio can even be used, where applicable, for communications.

Other short- to mid-term ICT deployment capability options include:

- Laptops, workstations, PDAs, faxes, printers, copiers, or mass printing;
- M/S Office, Adobe, Collaboration tools, GIS, GPS, mapping, VTC, and analysis and visualization tools, WiFi, language translators, VoIP;
- LANs, routers, servers, wireless coms, digital cameras, GPS receivers;
- power sources (generators, batteries, solar);
- Cell phones, satellite phones, VSAT, WiFi tower, and COWs;
- Prearranged satellite access, ISPs, and Teleport use;
- HF, VHF (including ground to air), and UHF radios and AM/FM transmitters;
- Land Mobile Radio Service (LMRS) for emergency services (police, fire, and rescue);
- Transmission and switching capabilities to support voice and data surge capacity needs;
- Kiosks with call centers and Internet cafes.

Mid-Term Reconstruction

Mid- to longer-term reconstruction options should include "leave-behind" infrastructure for local partners, whether they be private, governmental, or nongovernmental agencies. Depending on local regulations and capability constraints, leave-behind infrastructure can be used to jump-start host-nation ICT infrastructure recovery and reconstruction and provide commercial opportunities to seed economic revitalization through ICT infrastructure development. To make a leave-behind strategy effective, a host-nation ICT strategy and action plan should be developed. Considerations to take into account include:

- Understanding the affected nation's information culture and market, for example, how do individuals obtain and use information?;
- Understanding ICT policies, regulations, and laws;
- Articulating a vision, such as developing a "knowledge society" or "regional software development hub";
- Composing a strategy to achieve the vision;
- Building national capacity in the areas of infrastructure, regulations, laws, management, and resources;
- Developing electronic commerce and e-government;
- Education and training;
- Financing; and
- Fostering development of an ICT industry.

With those considerations kept in mind, planners can review their "toolkit" options for a graduated, mid-term ICT deployment capability, which may include:

- Expanded LMRS for emergency services;
- COWs plus fixed sites and WiFi and WiMax hot spots;
- VHF and UHF repeaters and relays;
- VSAT networks;
- AM/FM radio and TV transmitters;
- Mobile switches to interface with PSTN (PABX-like); and
- Kiosks, call centers, and Internet cafes.

Long-Term Leave-Behind Support

Long-term, leave-behind support should contribute toward the growth of functioning telecommunications and computing sectors, empowering a stable host government to identify goals for communications and IT development. Among the types of infrastructure that governments likely will seek to build are:

- Wireless networks
  - Mobile voice
  - Mobile data
- Broadcasting networks
  - Local, regional, national voice/data
  - AM/FM radio
  - Television
- Internet backbone and access networks
- Public switched telephone network (PSTN) reconstruction and modernization
  - Local, regional, national, international
- Terrestrial broadband transmission networks
  - Cable, fiber, line-of-sight wireless
- Satellite networks
  - VSAT, earth stations, teleports, hubs

The following guidelines address the issues involved in setting up ICT systems and capabilities, operating those networks and systems, and then departing, leaving behind the seeds of redevelopment:

- Get in very fast when an emergency develops;
- Offer high-quality, open, and multi-model communications;
- Facilitate improved cooperation among emergency responders, the host nation, and global experts;
- Serve as a model of flexibility, accessibility, interoperability, and ease of use;
- Ensure the presence of cultural insight through both training and guidance;
- Bring in the best automated translation tools and human translators that can be found;
- Cost nothing to the affected area;
- Spend a little for the local economy;
- Transfer and leave behind whatever appropriate infrastructure host nations wish to receive;
- Train the local technicians and experts who will continue to support that infrastructure;
- Maintain long-term support links; and
- Depart as soon as it is appropriate, advisable, and feasible to do so.

Data and Information Management

In S&R and HADR operations, the participants deal with data that are by nature multi-sectoral, multi-dimensional, contextual (historical, cultural, and ethical), multi-source, and non-standardized. In addition, they must confront the challenges of data overload while simultaneously redressing data gaps. Consequently, one of the major problems is synthesizing this mass of heterogeneous data into a form that can readily be digested by decisionmakers.

Timely and accurate information is integral to civil-military actions in natural disasters, post-conflict stabilization environments, and complex emergencies. The international civil-military community's ability to collect, analyze, disseminate, and act on key information is fundamental to an effective response. Better information, leading to improved responses, directly benefits affected populations. Over time, improved assessment of impacts and responses through better data collection and management contributes to a more complete global database on disaster impacts, leading to better risk assessment and prevention and preparedness activities.

Data Management

Given the current operational challenges facing the participants in real-world operations, it is clear that data management is increasingly an issue. During the early 2000s, numerous workshops were held on the subject of data for S&R and humanitarian operations. Table 1 is a partial list of the exceedingly challenging data issues identified during recent workshops.

Table 1. Selected Data Issues		
Data sharing	Data	acquisition
Data conversion	Data	reuse
Lack of good data dictionaries		Data bloat
Lack of knowledge of original purpose		Data protection
Data subrogation	Data	naming conventions
Data purity	Data	maintenance
Metadata policy (e.g., standardization)		Data shelf life
Ontological development for intelligent searches		Data reconciliation

It is widely recognized that many barriers impede the effective reuse of data. These include:

- A lack of knowledge about the existence of legacy data;
- Security or proprietary ownership restrictions;
- The quality of metadata (that is, the failure to document conditions of collection);
- Variance in definitions, language, and measurement instruments;
- The form of accessible data;

- Rapid changes of technical data, requiring frequent updates; and
- The fear that data could be misused, misunderstood, or lead to adverse consequences.

The goals of the S&R and HADR communities in regard to data can be broadly divided into two macro-objectives. First, it is critical that the data be *available* to the user. It must be visible, accessible, and rapidly accessible. Second, the data must be *usable* by the recipient. This implies that it must be understandable, trusted, interoperable, in the appropriate format, and responsive to user needs.

As a foundation for trying to improve the situation, it is necessary to develop the broad data needs and information exchange requirements for the S&R/HADR *Community of Interest* (COI). Second, it is appropriate to formulate an architecture to guide the activity. This would include a characterization of the "as is" situation, the desired "to be" situation, and the path to evolve from the baseline. Third, it is vital to rapidly develop a core capability to satisfy immediate needs. Finally, it is important to commit to incremental development that reflects feedback from the users, provides additional capabilities and functionalities, and incorporates emerging information technologies.

To improve the current situation, several steps must be taken. First, to ensure that data are visible, available, and usable when needed, they should be "tagged" with metadata indicators to enable discovery by users. All data can then be posted to shared spaces to provide access to all users—except when limited by security, policy, or regulations. Instead of defining interoperability through point-to-point interfaces, the community will then progress to enabling the "many-to-many" exchanges typical of a "network-centric" data environment.

Planners can formulate a data vision that is predicated on three key elements:

1. *COIs* to address organization and maintenance of data.
2. *Metadata*, which provides a way to describe data assets and the use of registries, catalogs, and shared spaces, which are mechanisms to store data and information about data.
3. *CIE Services* that enable data tagging, sharing, searching, and retrieving.

To institutionalize data management, the COI must govern data processes with sustained leadership. It must actively incorporate data approaches into COI processes and practices. The COI also must advocate, train, and educate personnel in data practices and adopt metrics and incentives. Meanwhile, data must be understandable. This can be accomplished by defining appropriate ontologies and metadata. The latter should subsume both content- and format-related metadata. For data to be trusted, it is necessary to associate data pedigrees and security metadata, and to identify authoritative data sources.

In supporting data interoperability, one should register metadata, associate format-related metadata, identify key interfaces between systems, and comply with agreed open-system interface standards. Being responsive to user needs implies involving users in COIs and establishing a process to enable user feedback.

To make additional substantive improvements in data support to S&R and HADR operations, it is vital that the civilian and military communities take several challenging steps. The most of important of these steps is an initiative to transform the culture of data from one of hoarding to one of sharing.

To do so, efforts must be made to dispel the fears that permeate both the S&R and HADR communities regarding the potential misuse or misunderstanding of data, and of the adverse consequences that could result. This initiative must be undertaken and sustained at the highest levels of leadership among the civilian and military participants. Establishing adequate metadata organization is the key technical issue. This is a challenging problem, and the S&R and HADR communities should begin to address it seriously and immediately.

Information Management[19]

Information management, which addresses how data are compiled, organized, and utilized faces its own ongoing challenges. Interested practitioners in the field met February 5-8, 2002, in Geneva to take stock of achievements in the humanitarian information management field, identify future challenges, and agree on next steps. Based on their collective experience, the participants in this international meeting identified a number of principles for humanitarian information management and exchange that apply not only to humanitarian activities but also to civil-military information management in general. In complex emergencies and natural disasters, the civil-military community should:

Define user needs and emphasize data sets and formats that directly support decisionmaking at the field level. Identify user groups, conduct user requirement analyses, inventory information resources, and define core information products based on user input. Develop and implement information products on operationally relevant themes, such as the location and condition of the affected population, as well as factors affecting access to affected populations. Use templates, such as the Rapid Village Assessment (RVA) tool to speed data collection. Create maps to effectively communicate information to decisionmakers.

Collect and analyze base data and information before and throughout an emergency. Gather, organize, and archive data and information on operationally relevant themes for high-risk areas in preparation for emergencies. Maintain and enhance data sets during emergency responses. Document and archive data so that it is easily accessible for future use.

Maintain and promote data and information standards. Follow generally accepted standards for information exchange, such as the Structured Humanitarian Assistance

---

[19] UN OCHA Symposium.

Reporting (SHARE) standard to promote data sourcing, dating, and geo-referencing. The SHARE standard facilitates integration of data from multiple sources and enhances verifiability, assessment, analysis, and accountability. Geo-referencing data during collection allows cartographic presentation and GIS analysis. Create metadata catalogs as part of a standard documentation process with hand-over procedures.

Maximize resources by expanding partnerships. Recognize that data and information are collected and managed by a variety of actors, including national governments, UN agencies, NGOs, the private sector and research institutions, and that the contributions of these providers are crucial. Pre-establish inter-agency agreements and relationships at the national and local levels. Establish an ongoing process of personal interaction to create partnerships for information management and exchange. Use distributed networks and neutral portal repositories to assist with information sharing and to promote linkages to avoid duplication of effort.

Engage local and national actors in information projects. Develop networks of local communities and national NGOs, civil society groups, and the private sector and address the issue of local participation as part of overall emergency planning, monitoring, and evaluation. Build and strengthen the national/local capacity in information management and exchange and promote the transfer and use of local knowledge.

Maintain preparedness "toolboxes" for online and offline distribution. These toolboxes provide guidelines and reference tools for the rapid deployment of information exchange facilities, for example Humanitarian Information Centers (HICs) or CMOCs, or the establishment of Web sites and databases under a variety of field conditions. Toolboxes should include data standards, operating procedures, training materials, database templates, and manuals.

Define an exit strategy. Develop a clear phase-out strategy, including transitioning to development activities and creating archiving systems to maintain access by current and future stakeholders after the project is closed.

Preserve institutional operational memory. Define and adhere to sound data and information management policies and techniques for handling large volumes of information. Document datasets with metadata. Maintain quality control and organizational learning to avoid the need to start from scratch with each emergency and to maintain the quality of information services during emergencies.

Establish field-based HICs. Design HICs as open-access physical locations, incorporate existing capacities, systems, and information management activities according to identified operational and decisionmaking demands.. HICs should serve as neutral brokers of humanitarian information, providing value-added products and beneficial services to the field-based humanitarian community. Encourage broad participation from local, national, and international actors to facilitate and support humanitarian response activities. Form partnerships with specialized agencies and sector experts to conduct sectoral surveys and analyses.

Use appropriate technology. Ensure that field information systems reach the broadest possible audience. Be aware of the limitations of technology, both inherent and as related to availability. For example, keep in mind that the Internet, while powerful, is not a panacea and can be ineffective as a distribution channel to and from remote areas. Consider making data products, particularly databases, available via e-mail, CD-ROM, and for local downloads. Recognize that local staff's ability to work with the technology is an important determinant of success. Technology should be easy to use and be accompanied by training for local staff.

Use open data formats and interoperable technologies. Use COTS technology and create all information products using open data formats and interoperable technologies.

Promote awareness and training. Conduct technology training sessions for non-technical humanitarian staff, particularly national staff members. Educate senior decisionmakers in humanitarian organizations about the purpose, strengths, and weaknesses of information management and exchange. Broaden participation in information projects among affected and at-risk populations.

Involve the private sector. Consider the efficiencies of contracting out information management and exchange functions to the private sector, especially local private interests, when cost-effective and appropriate. Encourage a constructive role for the private sector by incorporating private-sector expertise into preparedness and planning activities.

Mobilize adequate resources. Include funding for field-level information management and exchange systems and projects in the overall resourcing of assistance programs.

Best Practices

The attempt to assemble "best practices" is, at its heart, an effort to distill clear guidelines from the real-world experiences of professionals who have, in effect, already faced the inevitable challenges and overcome them. In short, best practice statements are the record of how those challenges were minimized and resolved. They can then be published and disseminated as trail markers to make sure those who follow in their footsteps do not have to traverse the same terrain through trial and error. Before embarking on more detailed lists and descriptions of best practices, which are found in the following sections, it may be useful to establish the following rules:

Ten Simple Rules for Civilian-Military Coordination:

1. Understand the other participants.
2. Respect their legitimate limitations.
3. Pay special attention to geographic and sectoral boundaries.
4. Leverage the shared interests of participants.
5. Build and carefully use networks.
6. Take the initiative in information sharing.
7. Have multiple, simple, reliable means of sharing information.
8. Encourage training and preparation for task sharing.
9. Provide the tools to facilitate joint planning.
10. Avoid public criticism of any participants.

The best practices in this chapter are derived from multiple sources, such as the UN, DOS, DOD, USAID, industry, professional workshops, and other subject-matter expert sources. The best practices are arranged in sub-sections according to subject matter.

Best Practices for Civil-Military Information Exchange

Organizational Issues

• Ensure that all reports compiled by responders have clear time and date stamps.
Reports about conditions on the ground are time-sensitive. To address this, the military uses standardized date-time groups to log reports, allowing military units to sort out multiple observations of the same event over time and to determine the most recent report. The absence of such procedures in the humanitarian community can result in misinterpretation of reports. This can lead to erroneous tactical decisions on where to route relief supplies, if, for example, a report arrives late and reflects out-of-date conditions and data.

• Be sure to confirm which metrics are being used and try to establish uniformity among responders.
Almost all reports provide measurements of some kind, whether of volume or distance. The military within a given nation will use standard measurements for each type of situation, and it may not explain or detail those metrics in each document. Moreover,

different agencies employ different measurement scales, for example, many U.S. civilian groups measures distance in miles, but the U.S. Army employs kilometers.

- Establish the variations in meaning for terms used by providers and, wherever possible, be conscious of different usages.

Terminology differs among various organizations; common terms used by the military and civilians, alike, often have different meanings. For example, the term *sector* means a geographic AOR for a military organization, while it denotes a functional area, such as water/sanitation, food or shelter, to most civilian humanitarian groups.

- Try to avoid depending on the military for data that is time-sensitive and has not been previously shared. State the information need in terms of the decision that must be made rather than the raw data being used to make the decision.

The urgent need of humanitarian groups for certain data may not be readily evident to the military, which often has its own set of priorities and information needs. Also, the military often responds better when it receives a clear reason why outside groups want information. Instead of asking ambiguously for satellite imagery, ask for a forecast of which local mountain passes are likely to become impassable due to impending weather.

- It is important to know how data was collected, and by whom, to judge whether it may be valuable or verifiable as a basis for action.

Information-gathering capabilities of participants differ. For instance, the military is more likely to be effective in gathering data about tangible, measurable things, such as the length of an airfield or the number of structures in a village with intact roofs. Military units seldom have the skills to make more than broad observations about issues, such as food security, access for ethnic minorities, or potential political conflicts.

- Pay attention to the effects of institutional cultures, as well as the informational and ethnic cultures of the host countries.

Cultural aspects can point up shortcomings in capability. In many militaries deployed outside of their own countries, direct interaction with the population is limited either by policy, regulation, or tactics. Effective interaction, when it is permitted, may be further limited by language, culture, or distrust. Junior soldiers will collect much of the information collected by the military from or about the population. Moreover, they often do their collections while on patrol, a mode that may not lead to casual revelations of information by the populace. Some militaries do have dedicated, trained personnel to interact with the population and conduct CA and CIMIC patrols to collect or verify data.

The Intelligence and CA/CIMIC sections of military forces are unlikely to have native-language speakers or personnel with direct experience in humanitarian assistance. In the case of the CA/CIMIC/CMO section there maybe a number of officers or soldiers who have had little or no formal training in this function. They will often defer to the humanitarian community on analytical issues.

If a CA/CIMIC channel for information exists between lower and higher levels of command this will be the primary channel for reporting information, addressing issues, and dissemination. There is a natural tendency not to place bad news into command channels. Furthermore, the amount of analysis done on humanitarian issues—and the amount of attention given to information requests by humanitarian groups—will often depend on the individual commander's interest in humanitarian matters.

- Try to leverage the logistical, transportation, and other capabilities and assets of military units to enhance humanitarian data collection.

The military may be well positioned to collect data on the population and humanitarian conditions in remote or inaccessible areas. For example, military medical personnel often conduct clinics in areas that humanitarians may not be able to reach. With some training, military medics could collect data on indicators of malnutrition among the patients they see.

Dissemination of information within multinational commands poses its own special kind of problems. Language and culture are obvious problems. However, differences in communications equipment and in national restrictions on the sharing of intelligence and information may also cause difficulties. The fact that information has been shared with the headquarters or one contingent of a multinational force is no guarantee that this information will reach all contingents.

- Military public affairs and media specialists can be used to reinforce advocacy issues and common themes not necessarily associated with either the military or the humanitarian community.

For example, public health, driving safety, respect for human rights, weather reports, and warnings can normally be broadcast over a number of channels without being associated with a particular source. On occasion, joint press conferences or media briefings may be appropriate. This may be the case when attempting to manage disinformation or particularly damaging rumors. On rare occasions, joint press releases might be issued. However, it is more likely that separate press releases properly timed will have better impact.

- Know how to get in contact with the right people in the most expeditious and appropriate manner to facilitate resolution of civil-military issues.

When possible, contact lists should include alternate points of contact and addresses where personnel can be reached during non-duty hours. The military operations center should be able to locate key military personnel in an emergency, and the radio operator in the UN mission and other IO/NGO operations should be able to do the same. But there is no substitute for direct access.

- Establish a formal process for the civilian humanitarian community to request information from military forces.

If a CMOC is set-up, or a humanitarian liaison officer has been assigned to the military headquarters, this officer may be the appropriate person to manage this process. An important element is the screening of the request, which is needed to ensure that the military forces understand what is being asked for and why. Blanket requests for data, such as overhead imagery of the country, will probably not be granted. In addition, the system must be able to audit and track requests, to gauge responsiveness and timeliness.

- <u>Remember that when the military reacts to an emergency, one of the first things they will do is restrict access to their critical facilities, such as the operations center. The civilian elements must know how to gain access in emergencies and must make these arrangements before the emergency.</u>

Emergency communications and liaison procedures are normally maintained by the security officers and may be included in disaster contingency plans. However, both civilian and military organizations will maintain alternative internal communications plans. Knowing the back-up systems and how they work in an emergency allows the civilian element to assist in the reestablishment of communications and coordination when systems are disrupted.

- <u>A civilian-military coordination arrangement must be prepared to quickly translate between military and civilian mapping systems.</u>

Logistically, standard map sheets, coordinate system, and location names differ between civilian and military elements. The military will work from a standard map or map series and a common coordinate system. The same is not true of the humanitarian community, especially the NGOs. Most military organizations use the Universal Transverse Mercator (UTM) grid system and maps in common scales, such as 1:25,000, 1:50,000, and 1:250,000. Most commercial GPS receivers, however, work in longitude and latitude coordinates.

In some cases, sophisticated GIS systems will handle translation, but the most reliable means is to have a transparent overlay with the alternative grid systems for the military map and the most common humanitarian maps. In addition, a consolidated list of place names, with alternative versions in the native languages and variants in spelling, must be maintained because most incident information received from local sources will reference neither the military or civil grid system.

- <u>Know something about the specific history between military and civilian individuals and groups/units.</u>

Recent experiences between people and groups will often influence the quality of relations between them more definitively than doctrine or policy. The same is often true regarding military experiences and relations with civilian populations and elites. It is worthwhile to research these informal histories to more effectively communicate and anticipate the reactions to both information and situations.

- <u>Minimize or eliminate any scenario in which rumors or misinformation about humanitarian or military participants can be originated or spread.</u>

The environment of an emergency is very conducive to rumors and misinformation. Rumors about the military will circulate within the civilian humanitarian community and vice-versa. Erroneous information can have a very damaging impact on civil-military relations and the mutual perceptions of both military and civilian actors. Systematically monitoring this phenomenon is normally the responsibility of public affairs officers in both the military and humanitarian organizations. Someone in the civil-military community must be prepared to assist in this effort by providing ready access to correct information and countering damaging false or distorted information.

Technical Issues

• <u>Before sharing facilities, weigh cost savings and expedience against the possibility that services will be disrupted if someone decides to target military communications infrastructure.</u>
Some infrastructure, such as VHF repeater towers, is expensive and ideal locations are limited. It may be possible for civilian and military elements to share such infrastructure.

• <u>If military forces have the technical capability and are willing to evaluate or repair communications or computer systems, this may be an area of low-visibility, indirect assistance that could benefit the humanitarian community.</u>
Furthermore, repair and maintenance of communications systems often requires technical skills that may be limited in the local economy and costly to import. If such arrangements are made, be careful not to advertise them too broadly and try to limit them to emergency repairs.

• <u>At a minimum, a memorandum of understanding (MOU) should be negotiated to ensure equitable access to radio frequencies and satellite resources.</u>
Control of the use of the electro-magnetic spectrum is the responsibility of the host-nation government. Specific frequencies are reserved for specific activities. Important frequencies for international communication, such as air traffic control, are established by international treaty. Other frequencies, such as those used for commercial radio transmission, are controlled by the government and licensed. In the absence of an effective governmental agency to deal with the monitoring and enforcement of these matters, the international military and civilian communications specialists must come to an agreement on these matters or communications can be severely disrupted, especially in emergencies where radio discipline may be weak.

Developing and Maintaining Partnerships

• <u>Participants in S&R and HADR operations can maximize resources by establishing partnerships.</u>
No one organization can collect all of the data and information needed during S&R and HADR operations. Partnerships encourage trust and commitment among stakeholders and allow information systems to remain objective, accountable, and focused on common rather than narrow interests. An ongoing process of personal interaction can create partnerships for information management and exchange. Recognize that data and information are collected and managed by a variety of actors, including national governments, UN agencies, NGOs, the private sector, and research institutions. The contributions of these providers are crucial.

• To decrease dependency, engage local and national actors in information projects and develop networks of local communities and national NGOs, civil society groups, and the private sector.
It is important to address the issue of local participation as part of overall emergency planning, monitoring, and evaluation. Build and strengthen the national/local capacity in information management and exchange and promote the transfer and use of local knowledge.

• Promote trust and transparency through IT linkages.
Use distributed networks and neutral portal repositories to assist with information sharing, and promote linkages to avoid duplication of effort.

• Partnering with the media can be an effective way of communicating information to the affected population.
Use local and international media outlets to inform the local population about humanitarian relief operations.

Establishing a CIE

Commercial ICT solutions exist to rapidly extend communications connectivity and capacity and information services into S&R and HADR areas of operation. The Internet has become the *de facto* civil-military collaboration environment, and both civilian and military elements employ Internet portals to facilitate information sharing. Data and information management in this ad hoc environment remains a challenge.

• Conducting an assessment of information needs and existing knowledge resources in advance is a first step in building a CIE.
This allows participants to identify the gaps in data, information, and knowledge. A standardized meta-data (source, date, geo-reference, definitions) approach can complement all collected and shared information, so that it can be pooled, compared, verified, mapped, and used for analysis.

• Establish and use collaboration networks to create communities of interest among individuals in multiple organizations.
Collaboration networks are a means to capture and share knowledge and dismantle organizational stovepipes. It is important to create an environment of willingness to share data and information. Civilian and military actors in S&R and HADR fields should work for an agreed strategy, CONOPS, systems architecture, and standards for ICT support.

• Develop an ICT technology roadmap for the near-, mid-, and long-term future.
Plans should address ICT capabilities (cellular, satellite phones, VSAT, wireless networks, PDAs), and plan to fill gaps and build capacity. Collaboration and information sharing (peer-to-peer, web pages, portals, GIS) should be accounted for and promoted. It is extremely important to create a metadata repository (registry, catalog, shared workspace).

The goal is to reach an agreement on organizational arrangements for creating and maintaining a civil-military CIE—and the "system of systems" supporting this environment. One result should be development and acquisition of ICT "fly away" packages for use by civilian government and military participants. These packages of equipment will facilitate collaboration and information sharing with IOs and NGOs. Moreover, they can be used as leave-behind ICT starter packages for building host-nation capacity.

Collecting Data, Managing Information, and Seeking Knowledge

Data, information, and knowledge are all different. In S&R and HADR responses and complex emergencies, information overload usually occurs in the form of reporting, but the systematic collection of standardized data that can be stored in retrievable databases is lacking. Documentation about the application of lessons learned and best practices in decisionmaking also falls short.

- <u>Humanitarian organizations need to establish strategies and systems for the management of data, information, and knowledge.</u>
These functions need to be planned for, resourced, and set up prior to—or in the early phases of—a crisis response effort, rather than being an afterthought. Training in these functions and systems should be held for humanitarian organizations' staffs in the field and at the headquarters level.

While some information needs concerning complex emergencies are common, other needs are more specific to different personnel within organizations throughout the humanitarian community. For example, U.S. government decisionmakers need "big picture snapshot" information to consult with other decisionmakers, negotiate with partners or adversaries, and inform Congress, the media, and the public about the emergency. They need to understand the issues, be aware of the assistance being provided, and be alerted to potential problems.

- <u>Establish procedures to supply and disseminate a common operational picture or situational awareness concerning a humanitarian emergency.</u>
Complex emergencies, S&R deployments, and HADR responses are characterized by conflicting and contradictory information and confusion of facts, terminology, and perceptions. A coherent situational picture needs to be presented so that the humanitarian issues and needs can be understood and addressed.

- <u>Information should be collected, organized, and disseminated in a manner that will benefit the population and polity of the affected country, not that of the outside responders.</u>
External humanitarian organizations need to recognize that they do not have all of the information and should not automatically assume they know what is best for the population of the country. The external organizations need to incorporate and use the information that comes from the affected populations, the indigenous groups, civil society, and the responsible government agencies. Information and training need to be provided to the local and national staff and entities to enable them to serve the needs of the affected population, prepare for a transition, and support long-term development goals.

The HICs established by the UN in the field have made progress in sharing operational data and information for the benefit of humanitarian aid workers and the local population. These neutral information centers provide products, services, and tools specifically for national and international aid workers. These products, services, and tools include GIS maps, contact and tracking databases, and standardized assessment forms.

• <u>New technologies, such as PDAs, GPS, GIS, and virtual collaboration tools, require advance training to be used effectively.</u>
Organizations and their senior management should encourage their personnel to use these new technologies and tools as part of their work, but their value must be demonstrated through the quality and timeliness of output. Advance training and organizational commitment are necessary for these data-collection tools to add maximum value.

• <u>Short, simple, standardized templates facilitate the collection of rapid assessment and programmatic data so that it can easily be transferred into databases for GIS and analysis applications.</u>
Create and disseminate templates for sectoral assessments and analyses, such as the RVA tool or the Microsoft/Save the Children assessment tool for PDA, to speed data collection. Create maps and graphic presentations to effectively communicate information to decisionmakers.

• <u>Valuable information needs to be disseminated, whenever possible, on unclassified and open-source platforms and channels.</u>
Within the U.S. government, almost all of the data and information about S&R, HADR, and complex emergency responses comes from open-source or unclassified sources. All too often, defense and intelligence agencies disseminate their information using classified platforms and channels because it is their usual means of reporting, even though the data and information does not need to be classified.

Satellite imagery that used to be automatically classified is increasingly being declassified and released for public use. Also, commercial satellite-imagery-derived products are increasingly being made available for use by humanitarian organizations. The National Geospatial Intelligence Agency (NGA) in the United States has established licensing agreements with commercial venders to release high-resolution imagery to external humanitarian organizations working with the U.S. government in emergency relief situations.

• <u>Define user needs and utilize data sets and formats that directly support decisionmaking at the field and headquarters levels.</u>
Identify user groups, conduct user requirement analyses, inventory information resources, and define core information products based on user input.

• <u>Develop and implement information products on operationally and strategically relevant themes.</u>
Focus on the location and condition of the affected population and the assessments of needs.

- Consult data providers as well as affected populations when designing information products.

Without direct feedback to and from the data providers and affected populations, community information collection strategies will fail. The humanitarian community, governments, and beneficiaries must be thought of as customers as well as data providers.

- Follow generally accepted standards for information exchange.

Use guides such as the U.S. Office of Foreign Disaster Assistance Field Operations Guide and the Structured Humanitarian Assistance Reporting (SHARE) standard to promote data sourcing, dating, and geo-referencing to allow cartographic presentation and GIS analysis.

- Coordinate methodologies for surveys and needs assessments.

These should be coordinated and designed to avoid bias, and should include training and documentation. Assessment methodologies and criteria should be explicitly stated.

- Use metadata.

Develop metadata documentation for each new product. Create metadata dictionaries and handover procedures as part of a standard, transparent documentation process.

Preparing For a Deployment

Organizations need to establish strategies and systems for the management of data, information, and knowledge. These functions need to be planned for, resourced, and set up prior to any deployment. Prior training for these functions and systems is required, and whenever possible, ICT capability packages should be setup and tested before deployment into the crisis area.

Information flow during a crisis can be crippled by lack of time, scant resources, isolated decisionmaking, limited information sources, and disrupted communications among participants, making it difficult to gather and process accurate data in a timely way. Moreover, information gathered during an emergency tends to be reactive and behind the curve. Hence, one of the most important aspects of information management and exchange is preparation.

The following two sections address some general factors to be considered when preparing to deploy into a crisis area and to select satellite phones or terminals.

An ICT Deployment Checklist

When preparing for participation in a crisis response operation, several factors need to be considered to improve situational awareness and identify ICT and other needs to support deployment. The following checklist represents some best-practice recommendations:

- Maintain preparedness "toolboxes" for online and offline distribution.
  - Toolboxes provide guidelines and reference tools for the rapid-deployment of ICT packages and/or the establishment of Web sites, intranets, and databases under a variety of field conditions.
  - Toolboxes should include data standards, operating procedures, training materials, database templates, and manuals.

- Develop surge capacities for rapid deployment.
  - Maintain rosters of experienced ICT professionals.
  - Formulate equipment caches.
  - Establish training and exercise programs.

- Preserve institutional operational memory.
  - Define and adhere to sound data and information management policies and techniques for handling large volumes of information.
  - Document datasets with metadata.
  - Maintain quality control and preserve organizational learning to avoid starting from scratch with each new emergency.

- Define an exit strategy.
  - Develop a clear, phase-out strategy, including transitioning to development activities and creating archiving systems to maintain access by current and future stakeholders after the project is closed.

- Involve the private sector and academia.
  - Consider the efficiencies of contracting technical functions to the private sector, especially local private interests, when cost-effective and appropriate.
  - Encourage a constructive role for the private sector and academia by incorporating their expertise into preparedness and planning activities.

- Mobilize adequate resources.
  - Include funding for field-level information management and exchange systems and projects in the overall resourcing of assistance programs.

- Promote awareness and training.
  - Conduct technology-training sessions for non-technical humanitarian staff, particularly national staff.
  - Educate senior decisionmakers in humanitarian organizations about the purpose, strengths, and weaknesses of information management and exchange.
  - Broaden participation in information projects among affected and at-risk populations.

- Develop contact lists
  - Lists should feature key humanitarian responders and local personnel, with phone numbers and email addresses.

- Review host-nation legal, regulatory, and institutional considerations:
  - Identify customs and border crossing requirements
    - Personal (passports, visa, medical)
    - Equipment/software (taxes/duties)
  - Review licensing policies regarding use of WiFi, radios, cellular, satellite phones, and satellite terminal landing rights (VSATs)
    - Spectrum usage constraints and local licensing
    - Language restrictions, for example Sudan local law only allows use of Arabic and English on open channels
  - Availability of a Status of Forces Agreement on treatment of foreign militaries
  - Implementation of Oslo Guidelines for the use of military assets in humanitarian operations
  - Host-nation capacity to respond including appropriate coordination structures and mechanisms
  - Status of infrastructure.

- Where possible, obtain host government permission to import and operate communication equipment within and across borders.
  - Normally, a radio spectrum license is granted by the ministry of telecommunications or, increasingly, from the independent telecom regulatory authority or agency. Obtain several frequencies for redundancy and expansion.
  - Clarify allocations and availability of frequency assignments.

- From the outset, keep the host government informed of requirements and work with them to ensure that frequencies and radio licenses are allocated to all the various participants.

- Identify ICT capabilities needed and available, such as:
  - HF, VHF, and UHF radios and radio rooms
  - Voice (VoIP, PBX, telephone, cellular—GSM and/or CDMA, satellite phones)
  - Printer, copier, fax, text messaging, data, video, digital camera, VTC, maps, and imagery
  - Internet (dial-up, teleport, networked), email, web sites, metadata repository, "cyber-café" connections
  - Computers (workstations, ruggedized laptops, PDAs), WiFi, fixed LANs, routers, large format monitor (incident mapping)
  - Word processing, mapping, and collaboration tools, GIS
  - Mobile cellular services and applications—SMS, picture, video
  - Satellite systems (Inmarsat and VSAT, satellite coverage footprints, GPS)

- A limited number of links to provide connectivity to some peripheral offices
- Power sources (car batteries, batteries, solar, multi-fuel combustion).
- Contact agencies that may already be there to:
  - Determine the type of communication equipment they are using (common frequencies can save a lot of time and effort).
  - Obtain equipment that is compatible with implementing partners.
  - Obtain the information and specifications needed to order radios. If possible, try to borrow radios for the duration of the emergency.

- Check if your organization has a blanket agreement for communications with the government.
  - Frequencies may already be available for use, obviating the need for further government approval.
  - Inform the government of your intention to join the radio network and advise officials what type of equipment you will be using.

- Make one person responsible (if no communications officer is available) for acquiring, installing, and operating the radio devices.
  - This person should inform everyone of the available frequencies, voice procedures, radio etiquette, calling schedules, and where channels are available.

- An important factor to consider for effective communication is the power supply. Is it reliable, sporadic, or non-existent? Consider using solar panels or a generator. The staff member assigned to communications set-up should determine a cost-effective and realistic scenario.

- Develop cultural awareness and civil-military situation awareness.

- Use a rapid response ICT assessment team (2-4 persons) in advance of full deployment to establish needs and conditions on the ground.

- Determine communications requirements:
  - Use a common approach wherever possible
  - Encourage all participants to involve themselves.
    - If other agencies have communication technicians, ask for their advice on implementing a universal system for all aspects of communication.
      - Show them the information gathered in your initial assessment of communication resources.
      - Determine, with the technicians, the best-case scenario and disseminate to all participants.
  - If security is an issue, each staff in the field should have a handheld radio
  - When deciding on your requirements,

- Plan for future staffing levels; don't allow activities to come to a halt because staff cannot be deployed due to lack of radios.

- Radio batteries are critical to the operation of your radios.
    - Modern radios use rechargeable nickel-cadmium batteries.
        - These batteries operate best and last longer when they are used (fully discharged) and then recharged.
        - Keeping the radio and battery in the charger and using the radio infrequently reduces battery effectiveness, especially for older-model radios.

Considerations for Selecting Satellite Phones or Terminals

Before investing in satellite phones or terminals, it is important to review operational needs and considerations, such as the desired geographic coverage area (global versus regional), as well as weight and mobility needs, whether the deployment calls for use of the phones in remote and rural areas, and whether usage calls for mobile or fixed installations. Finally, consider what performance is required, in terms of cost, quality of service, and reliability. It is also necessary to make an initial assessment of the operational area's local telephone and mobile service, as well as local Internet access, to determine whether these local assets can satisfy identified operational needs.

Current commercial satellite access capabilities range from low to medium bit-rate portable terminals such as Iridium (2.4 kilobits per second), Mini-M (2.4 kbps), Thuraya (9.6 kbps), Globalstar (9.6 kbps), M4 (9.6 to 64 kbps), and BGAN/RBGAN (64 to 144 kbps) to broadband mobile and fixed installation VSAT satellite terminals that can operate in the range of 512 kbps up to 8 megabits per second and more, depending on satellite dish size and other system design factors.

The following are some simple considerations to help determine whether a portable, mobile or fixed installation satellite terminal option is the appropriate arrangement to consider for meeting operational needs.

Portable Terminal Considerations

If portable, worldwide coverage is needed, the Iridium satellite phone is an option to consider. The Iridium phone is used primarily for voice communications, but it also works well for email. Iridium can be used in remote and rural areas within a region and, because of the satellites' Earth coverage, can also be used if to travel to different regions of the world is necessary.

In the Middle East and much of Asia and Africa, the Thuraya handheld satellite phone service has a proven performance history for voice communications. The Globalstar service may also be suitable in areas other than Africa. In the area of interest, GSM cell phone service (up to 9.6 kbps) may be used, as well, but be aware that cellular coverage in most parts of the world is reliable only in cities. Remember, also, that the Iridium,

Thuraya, Globalstar, and Mini-M satellite phones, and GSM phones can be used to send and receive emails.

If the requirement is primarily email, with occasional activity on the Web, the Regional Broadband Global Area Network (RBGAN) terminal (up to 144 kbps) is the preferred solution. In some cases, it may be necessary to install an external antenna for the RBGAN, with at least a 12-meter cable, to get the desired performance. Since it can be easily carried to the field, the RBGAN is well suited for travelers who must send and receive email regularly—not just on occasional visits to an office or Internet café. The M4 terminal (up to 64 kbps) is also highly portable and can be used for rapid setup of Internet access.

A wide range of phones and terminals are available to choose from. Selection of any of these options is usually driven by cost and intended usage.

Mobile and Fixed Solutions Considerations

If it is necessary to access Web sites regularly or send and receive large amounts of data, a broadband VSAT solution is needed. VSAT is a satellite communications technology that uses a small earth antenna, usually 1-2 meters in diameter, for portable and mobile configurations. The smaller terminals tend to support medium bit rates up to 512 kbps, but a number of rapidly deployable 1-meter-dish, auto-acquire, flyaway VSAT packages are now available, offering up to 1.5 mbps data rates and Internet access. The larger-antenna fixed installations offer up to 8 mbps or more of broadband data service. Satellite service providers offer even higher capacity system options.

Acquiring a VSAT system is not the same as purchasing DSL or cable services—nor is the performance comparable in many cases. It will, therefore, be important to carefully identify user functionality and performance needs, the physical and operational environment considerations, and cost constraints. A cost versus performance trade-off should be conducted to select the desired arrangement to meet operational performance within cost restrictions.

VSAT terminals, networks, and services can be acquired and set up by the responder or leased from a VSAT service provider—a company or organization, such as an NGO, providing the data service over a commercial, satellite-based network. The service provider can be either a satellite operator directly, or a system integrator. A satellite operator usually owns the satellites in its network, as well as the ground stations (teleports). A system integrator is a company that does not own its own satellites, but makes use of satellites and teleports owned by many different satellite operators. This can be advantageous because integrators can select from different companies' capabilities and coverage options rather than be locked into one operator's satellite fleet. The service provider also provides the needed physical and information service capabilities within the operational area (VSAT terminals, LANs, WiFi, tents, tables, chairs, computers, routers, and other needed parts), and associated network administration support arrangements.

The VSAT solution can provide e-mail transmission and synchronization, Web access for surfing the Internet, and telephone access—either directly or over the Internet using VoIP applications. The VSAT terminal in the operational area is typically connected to a fixed and/or wireless LAN that serves a number of subscribers. Within a region, more complex arrangements can be found, such as multiple VSAT terminals interconnected to form a network serving multiple users in different geographic areas.

Considerations for Selecting VSAT Systems and Services[20]

A number of factors need to be considered when selecting a VSAT system or service. These factors are both operational—relating to how the companies do business and provide the products, service, and support—and technical, involving how the network and systems are designed and operated. Other factors include user needs, for example, what combination of voice, data, or Internet access is required, and geographical and environmental considerations. These typically involve whether the satellite to be used is too low on the horizon or obscured by mountains, tall buildings, or trees—or whether the VSAT is to be installed in areas of high rainfall, which can attenuate the signal. It is also important to ensure that the provider is using the hardware platform in a manner consistent with good reliable performance. The service provided will only be as good as the platform, and vice versa.

Operational, Technical, and Other Considerations

In purchasing a satellite system or service to meet operational performance needs, it is important to understand end user needs and how the network will be used, the operational environment, service and system provider capabilities, and system cost versus performance trade-offs. Some factors to be considered include, but are not limited to, the following:

Operational Issues

1) Coverage area. Does the service provider offer coverage of the area you want to work in? If you wish to provide connectivity to multiple offices in a region or worldwide, does the provider have coverage in all the areas, or will you need several providers to meet the needs of all your sites?

2) Large satellite fleet for coverage options. Several satellite operators have large enough fleets to provide good worldwide coverage. Another very valid option would be to use a service integrator that may provide service on different satellite networks, depending on coverage and footprints. This is more of a consideration for an organization that plans to support multiple, geographically diverse locations and wants the benefits of working with a single service provider. Another related factor is satellite failure. If a satellite fails, the company with more resources can move service or sometimes even maneuver an in-orbit spare satellite into position to provide ongoing service.

---

[20] DRASTIC, Global Satellite Broadband Data Service: Considerations for selection of a VSAT service and hardware platform, www.drasticom.net, accessed July 11, 2005.

3) Number of service providers required. Economies of scale on costs and quality of service and support can be obtained through network size. To that end, it may be better to select a service provider that can service all your sites in as large a region as possible. Having to interface with a single company also reduces staff workload at your headquarters office.

4) Network size economies of scale. Is the network you are buying service on optimized for the best economies of scale that can be passed on to the customer? Many integrators buy bandwidth and equipment in bulk but do not pass the savings on to the end customer buying in single-lot quantities. Another benefit of using a larger service provider or integrator is that the economies of scale extend to supply, support, and training, to name just a few factors.

5) Satellite operator vs. service integrator. Many integrators also provide equipment, and in some cases, their prices on equipment and monthly service may be cheaper than buying directly from the satellite companies and manufactures. The greater savings comes from their buying in great volume and passing the prices on to clients.

6) Data security at hub and hub countries. The satellite signals from a geographic region are all connected to the Internet through hub equipment, which is located at a teleport, or earth station. A single teleport can feed whatever satellites are in view. The hub is connected to a Network Operation Center (NOC), possibly located halfway around the world, and then on to the Internet. Some providers may use teleports or a NOC in countries where the security of the data flowing through those hubs may not be very high. One should examine the political and economic factors in the hub locations and possibly rule out certain high-risk countries.

7) Physical security at NOCs and teleport. Similarly, physical security at the NOC and the teleport should be a consideration. All facilities should be fenced. Access to the networking operational and management areas should be through keyed entry areas, accessible only to authorized personnel.

8) Company size and longevity. How large is the service provider company? Size is no guarantee of stability, but longevity is important in what is a moderately volatile industry. On the other hand, a company that is very large and established can become arrogant and unresponsive to individual customer needs.

9) Company financial solidity. Examine the history and financial viability of the service provider. The satellite industry is quite volatile. In an effort to provide low-cost services, some providers have priced themselves into bankruptcy. The chosen service provider should have a varied and large portfolio of products.

10) Installations. The installation of high performance VSAT systems is complex. The better companies recognize this, and require that installers be certified to industry standards. This helps ensure that installations will be accomplished in a manner that prevents the VSAT system from suffering interference, or from causing interference to other satellite users. Most service providers charge high prices for remote overseas installations: typically $7,000-$11,000 per site. Does the provider have certified staff in the country or region where the system will be installed, or do they have to send installers a long distance, at high expense?

11) Regional support. Similarly, does the service provider have local or regional support staff? In the unlikely event of equipment failure, the cost of having to fly a technician in from an overseas office could be expensive.

12) <u>Support hours of operation.</u> The service provider should have one point of contact for customer service and support, available around the clock. Anyone calling the help desk for support must be treated professionally and quickly. Some providers do not staff their support systems over the weekends or at night, and they can take several days to respond to a problem.

13) <u>Proactive support and monitoring.</u> The performance levels on some services and hardware platforms can be monitored continually by automatic software at the NOC. This is not monitoring of the content of data transferred, but rather keeping an eye on the remote site hardware to make sure it is performing properly. If the monitoring software notes a failure or even a downward trend over time, it flags that site for attention. In most cases, the NOC can dispatch a technician to investigate locally and even fix the problem before a part fails completely. This capability is commonly found on professional grade equipment and service, and seldom noted on consumer grade services.

14) <u>Levels of support.</u> The service provider needs to have a documented support escalation process and procedures with defined timelines. The process needs to allow for the provider to have direct access to the satellite operator and the hardware manufacturer's technical support staff on a 24/7 basis.

15) <u>Responsiveness.</u> The service provider must be very responsive to phone calls or emails. Staffs need to return calls or emails within several hours, or alternate contact people should be available.

16) <u>Landing Rights.</u> The service provider must be willing to assist in cases where a government has not yet granted landing rights for the satellite in question. Landing rights are permission from a government to a satellite operator granting them approval to provide service within that country.

Technical Considerations

1) <u>Frequency band of service.</u> Satellite service is provided on two frequency bands: C-band and Ku-band. Ku-band requires a smaller dish and is less expensive to install. The signal only illuminates a small region and is commonly used in situations where a large number of satellite users are concentrated. C-band service can cover whole continents and is available pretty much everywhere. It also works much better in high-rainfall areas.

> a) <u>Ku spot beams for high demand/population regions.</u> Ku band is preferred, where practical, due to smaller dish size. Smaller 1.2 meter dishes are cheaper to ship and install. They are less visible in the community (sometimes a good idea for security reasons). However, in some areas, a 1.8m dish is still required because of weak satellite signal strength. Ku-band also suffers from rain-fade in high rainfall areas where rain attenuates the signals.

> b) <u>C-band global and hemispherical beams for lower population density and tropical areas.</u> Ku-band service is not as universally available as C-band on a global basis. C-band service does cover all regions, including low population density regions. C-band is also much less affected by rain fade, making it much more usable in high rainfall areas, especially the tropics. The downside is that C-band usually requires a 1.8 or 2.4m dish, and the equipment costs more to purchase, ship, and install.

2) <u>Proxy caching at the hub.</u> Proxy caching at the hub can speed up web access for clients. This becomes less of a requirement if the connection to the Internet at the hub is very fast.

3) <u>Web content filtering at the hub.</u> Many users are concerned about the material readily available on the Internet. They are concerned about immorality, violence, and hate content negatively impacting the morals of their users. This is an especially strong factor to the governments in the Middle East. In addition, much of the malware such as adware, spyware, and viruses that infect computers come from these web sites. Even inadvertently opening one of these web sites for a couple of seconds can be long enough for a "drive-by" installation of unwanted software. Once infected, the compromised computer can very quickly generate enough traffic to cripple the satellite connection from the remote site and, in severe cases, seriously affect other remote sites on the satellite network. In those cases, the satellite operations center will sometimes shut down the offending site to protect the throughput of all the other network remote sites. A Web content filter at the satellite hub provides the best solution. This greatly reduces the risk of computers becoming infected from websites and reduces the requirement for active filtering to be done at each remote site.

4) <u>Quality of Service control.</u> Internet traffic can be either "bursty" in nature, such as Web surfing or file transfers, or it can be streaming, as with telephone voice calls and audio and video feeds. "Bursty" modes can tolerate several seconds of delay when a satellite channel is very busy. Non-bursty, real-time data, such as voice transmissions cannot tolerate such delays without resulting in choppy or garbled audio. Quality of Service (QoS) priority can be assigned to certain protocols, such as VoIP, giving time-critical voice traffic priority over File Transfer Protocol (FTP) downloads. Multiple levels of priority can be assigned to each protocol, depending on its importance for the site.

5) <u>Committed Information Rate (CIR).</u> In addition, some networks allow for a portion of the assigned bandwidth to be committed to each site. This is called CIR service. The most critical traffic, such as VoIP phone calls, are assigned with top QoS priority and placed in the CIR bandwidth.

6) <u>Redundant high-speed connections to Internet backbones.</u> The satellite hub sites should have a minimum of two redundant high-speed (DS3) connections to Internet backbone facilities. Links from the primary to the backup must be invisible to remote client sites.

7) <u>Backup Power at the NOC and teleports.</u> Every site that processes the data between the Internet and the remote site should have backup power sufficient to operate for at least three days if commercial power fails. This is typically met by using a large uninterruptible power supply (UPS) and an auto start generator. The NOC and teleport operators must maintain an adequate fuel supply on site.

8) <u>VoIP support.</u> The service provider must, at minimum, pass VoIP traffic to the Internet. Individual customers can use either their own gateways or a public gateway. The service provider may make a VoIP gateway service available. If so, it should offer various billing models, including prepaid scratch-cards or the equivalent.

9) <u>Ability to rapidly change bandwidth allocation.</u> The service provider should be able to increase bandwidth allocation quickly and readily without charging a fee for changes.

10) <u>Integrated webcasting IP delivery capability for distance education.</u> The service provider should be able to provide webcasting (delivery of data to many sites) for

distance education and conferencing. This could be live web casting or a service where data is pushed out to remote sites and cached locally for later use.

11) Mail relay. The service provider needs to provide a mail relay server for POP3/SMTP users to send mail. Mail servers increasingly do not permit mail to be sent from sites they cannot verify, which is frequently the case with VSAT systems.

Bandwidth Issues

1) Single Carrier Per Channel (SCPC) vs. Shared. A satellite in geo-stationary orbit relays data between the teleport and the remote site. Each satellite has many transponders, which are repeaters capable of relaying a tremendous amount of bandwidth, typically on the order of 25-50 mbps. That is far more than an average remote site will need. Leasing an entire transponder can cost $100,000-$150,000 per month. Consequently, service providers divide up the transponder's capacity among many different carriers, with a variety of bandwidths available.

    a) SCPC. Even the bandwidth for a relatively small channel can cost $3,000-$5,000 per month because the channel must be paid for whether you are transferring data or not. The higher cost is validated if the remote site needs to always have all of its bandwidth available and ready to use at any time.

    b) Shared bandwidth. Most operations, however, do not need the full amount of bandwidth at all times. A very cost-effective way to reduce the bandwidth requirement is to share slices of time on the transponder. This is a Time Division Multiple Access (TDMA), or shared, service. If 10 remote sites share a carrier, then the over-subscription ratio is 10:1, and each site pays approximately 1/10 the cost of the equivalent SCPC circuit.

2) Over-subscription ratio. How well a TDMA shared service works depends a great deal on how many remote sites are on a given carrier. Top-level corporate or Internet café users should have no less than a 4:1 over-subscription. Medium offices do very well at 10:1, and small offices are fine at 20:1 service. Home users with just a single PC might do fine at 30:1. But beware of unscrupulous service providers with very low-cost service aimed at consumer services that run at 50:1 or even 80:1 over-subscription. This level of service will very often be so slow as to be unusable for Internet service, especially VoIP calls. The situation is even worse if the remote sites, instead of having a single PC, have 10, 20, or more computers in an Internet Café at the site.

3) Burstable CIR service. A shared TDMA carrier works well for bursty Internet web browsing activity, but it does not do well with VoIP phone and other real-time voice traffic. These require some guaranteed bandwidth, like SCPC, to work well and deliver reliable voice call capability. An SCPC channel is all CIR. A portion of the shared bandwidth should be committed to each site through CIR service. Most sites just need a small portion, which can be provided. This type of service is a shared burstable CIR service, and it reflects the best value and performance for most users.

4) Jitter. This is a measure of how even the latency is over the satellite and how long the signal takes to get from one side to the other. It is imperative that the latency remain fairly constant if the VoIP voice quality is to be any good at all. Networks with wildly varying latencies, typically encountered on lower-grade, best-effort systems, can all

but kill VoIP voice traffic. This is very commonly encountered on excessively over-subscribed services.

5) Best-effort or guaranteed performance. Most of the low-priced services available are "best effort" services. Throughput or performance are not guaranteed. At the end of the day, even if the channel becomes so congested as to be unusable for hours every day, the service provider can simply say "that is the best the network can do." Often, the advertised data rates are theoretical maximums, not typical throughput. One differentiator between consumer-grade, best-effort systems and professional-grade services and equipment is that the latter operate under a "service level agreement" that guarantees performance and reliability.

6) Service Level Agreement (SLA). An SLA defines a minimum performance level and guarantees of network availability, latency, and packet loss. Typical SLA features include:

a) The network will be available to the customer an average of 99.7 percent of the time per calendar month, except for standard service events, and for events outside the carrier's control or at the remote site.

b) Packet loss will not exceed 2 percent during a calendar month.

c) Latency shall not exceed 750 mSec for a single satellite hop.

d) Jitter must be below 10 percent to maintain good VOIP connections.

e) CIR speeds shall be met at all times.

7) Data rate definition. Different service providers define the bandwidth of their services in different ways. When comparing one service to another, be sure to convert values to the same parameters. Information Rate and Transmission Rate are the most frequently quoted to represent the maximum theoretical throughput of the channel. They do not address typical throughput. Request in writing what typical throughput will be. The best way to measure and have bandwidth quoted is by IP throughput. This is a much more realistic measurement of the performance of a channel.

8) Non-pre-emptable service. The service bandwidth can be either pre-emptable or non-pre-emptable. This becomes a factor if the satellite transponders all become full, or inoperative. In these cases, a satellite operator or service provider may have to move some or all the users to a new satellite. Priority is given to customers with non-pre-emptable service. In case of a partial failure, non-pre-emptable customers will often be allowed to stay on the existing satellite service.

Equipment Non-technical Considerations

1) Manufacturer size and longevity. The equipment manufacturers should be moderately large and well-established, with a sufficient history to offer some expectation of future longevity. This is true for manufacturers of modems, outdoor units, and antennas.

2) Company financial solidity. The chosen equipment suppliers should be financially stable and not undergoing financial reorganization or bankruptcy. One should be reasonably assured of future longevity.

3) Equipment customer support. The modem manufacturer's support staff should be available 24 hours a day, every day of the year. Ask if the manufacturer has a published technical support escalation process. End users or field installers and

technicians will seldom need to contact the manufacturer directly, but it is important that the service provider have direct access to the manufacturer's tech support staff as the last level of escalation on technical issues.

4) Age of equipment design. The platform chosen needs to be relatively recent in design, and not an earlier design approaching the end of its useful and competitive life.

5) Upgradability. The hardware platform needs to be upgradeable in the future as new firmware is released. Firmware updates should be automatically uploaded to remote terminals.

Equipment Technical Considerations

1) Bandwidth efficiency. The most expensive component of a VSAT system is the recurring monthly satellite bandwidth cost. The modem platform should be selected on the basis of optimizing the use of the available bandwidth. TCP/IP throughput should be at least 1 kbps per MHz of transponder bandwidth. Different modem technologies can vary by more than 30 percent, so this is a very significant factor.

2) Carrier Spacing. Guard bands also affect bandwidth efficiency on the satellite transponder. Each signal on the satellite tends to bleed over to the adjacent channel. To avoid interference, satellite operators leave a "guard band" in between adjacent channels. Some modems and modulation methods generate signals with steeper drop-offs on the edge of the signal than others. The more contained the actual signal is, the steeper the drop off on each side. This results in a smaller required guard band. This directly translates into less actual spectrum used on the satellite, and therefore a reduced cost. Most modems require a guard band of 20 percent of the channel width on each side. Ultimately, the user pays for the extra 40 percent of bandwidth. Some modems require as little as a 10 percent guard band on each side, a savings of 20 percent overall on the bandwidth and thus the cost of service.

3) Non-contiguous bandwidth. Traditional satellite modems require a contiguous section of the satellite transponder spectrum. As multiple carriers come and go over time, the satellite transponder spectrum can become fragmented. Some modems can use small, leftover slivers of bandwidth cumulatively. These small "leftovers" of bandwidth are often available at fire-sale prices to the service provider, who can pass the savings on to customers.

4) Forward Error Correction (FEC). All VSAT modems use FEC for transmitting data. FEC is a means of improving the Bit-Error Rate (BER) on both directions of the satellite link. Older modems and the current Direct Video Broadcast-Return Channel by Satellite (DVB-RCS) standard use the older Reed Solomon Viterbi (RSV) form of FEC. This method is very inefficient when compared to modem Turbo Codes (TC) or Turbo Product Codes (TPC). This inefficiency results in higher bandwidth and power required for the same BER as TC or TPC.

5) Shared service with CIR foundation. Some modem hardware platforms and service models are designed around shared service, but with a guaranteed minimum committed information rate. This is the best mode of operation in that it

ensures that no one site gets all the bandwidth, but that all sites get their minimum paid for bandwidth, and excess capacity not being used by other sites is available for "bursting" to the maximum carrier speed.

6) Bandwidth on demand - Rapid Allocation. The modem platform needs to reallocate burstable bandwidth very rapidly. Some hardware platforms assign burstable bandwidth to a requesting site for 10 or 20 seconds, even though they may just need it for a few seconds. Other remotes that request some bandwidth they are entitled to cannot get access to the previously allocated bandwidth until the other remote releases it. The more rapidly a system assesses and reallocates bandwidth, the more efficient the network is and the better the user experience. Some hardware platforms can reallocate at up to eight times per second.

7) Application QoS for VOIP and other protocols. The modem should be able to assign QoS priority to time sensitive applications such as VoIP and email. This will ensure high quality voice calls and data throughput and reliable email transport.

8) Queue determined bandwidth allocation. QoS should also prioritize bandwidth allocation based on the queue of data at remote sites. The site that has more data queued should be able to receive a greater portion of the shared bandwidth to reduce its queue more quickly.

9) TCP Acceleration (Pure IP over Air) in both directions. This is extremely important for effective web browsing. The verification process within the TCP/IP protocol works poorly over the high latency of satellite connections. The problem is specifically with the TCP protocol. TCP satellite acceleration involves removing the TCP data from the TCP/IP frame, replacing it with a protocol designed for satellite latency path. The TCP header is added at the other side of the satellite link. This process drastically increases data throughput over a satellite link. This can be accomplished with TCP accelerators at both ends of the link to obtain speed enhancements at both ends. These products typically cost $20,000 at the hub and $4,500 or more per remote terminal. A lower cost option is to use one-way acceleration of only the outbound data, but this approach requires the use of TCP acceleration software on every client PC at the remote site. The ideal modem should incorporate bi-directional TCP acceleration within the modem.

10) TCP Session Initiation Acceleration. When a new web page is opened, each and every element opens in a new session. TCP sessions use slow start up for each session. This is especially troublesome when opening a web page that has multiple links for content or many elements. Each one of these items has to do a connection/acknowledgement process sequentially. A modem platform that can bypass the need for end-to-end acknowledgements over the satellite link can greatly speed up this process.

11) Virtual Private Network (VPN) support. The modem platform and service provider need to pass all VPN traffic. Some VPN protocols work better than others over high-latency paths as found on satellite links. TCP packets embedded inside VPN data cannot be compressed because the VPN TCP headers are hidden inside the VPN frame. This will make VPN traffic run at uncompressed speeds. The solution is to use VPN accelerators specifically designed for VSAT applications. Some modem manufactures offer products specifically optimized to

work on a satellite path using their modem technology, allowing the benefit of the TCP compression to be regained even over VPN circuits. It is highly recommended that the modem/accelerator package be considered if a significant amount of VPN traffic is anticipated.

12) <u>Web compressions.</u> Web traffic can be very slow if a lot of detailed graphics are sent over the VSAT link. One solution is to use Web compression to compress the images and other compressible data through lossy compression prior to transmission to the remote sites. Some modem platforms include this capability, resulting in a further level of speed increase for Web browsing. The uncompressed images are available by clicking on the transmitted image.

13) <u>Integrated webcasting IP delivery capability.</u> The hub and remote platform should be able to support multicasting, the broadcasting of IP data from the hub to many sites. This is useful for distance education and conferencing.

14) <u>Built in Router/NAT/DHCP/DNS caching.</u> Care must be taken with this option. The VSAT system is just one link in the connectivity chain. Additional functions are needed at the remote site to build an efficient and workable communications tool. The VSAT system provides basic connectivity to the Internet. The local connection is output on an Ethernet connector. To provide service to a LAN, further networking functionality is needed locally. This includes a Router providing NAT, DHCP, and network security from the satellite WAN. Further functionality should include DNS caching and TCP acceleration. Combined, these items can easily cost more than $6,000. Some modems provide all this functionality within the modem, reducing system complexity, size, and power requirements.

15) <u>Built-in Web Caching.</u> To reduce the volume of traffic on the satellite link, a local web cache at the remote VSAT site can be used to cache frequently requested information. Although an external PC can provide this, some modems incorporate this function as an optional hard drive directly in the modem. This reduces complexity, cost, power, and maintenance requirements.

16) <u>Built-in 3DES satellite link encryption.</u> The satellite link is much more stable than a wired or wireless LAN. Remote sites may want improved data security, and the modem platform should be able to encrypt the satellite data path. This encryption should be industry standard, preferably 3DES, and add less than one percent overhead to the data.

17) <u>Automatic uplink Power Control.</u> Rainfall affects satellite performance on both the downlink and uplink of the satellite path. Uplink power is set quite tightly to be just enough to get a reliable signal to the satellite without overloading it. Rainfall will attenuate the signal, and in the case of a heavy rainstorm, the uplink signal may drop below the receiver threshold on the satellite, breaking the connection. The hub in some hardware platforms continuously monitors the signal strength received from remote sites. When a drop is sensed from rain fade, the hub remotely commands the remote site to increase transmit power to maintain the connection. More advanced modems go even further by varying the FEC coding under extreme signal degradation. This means that the satellite link will stay at full speed as long as possible under degrading conditions, such as a building rainstorm. When the rain gets extreme, rather than just dropping the link

completely, the coding is changed and the link gets progressively slower but does not stop completely.

18) <u>Remote Modem Disabling.</u> Should the satellite equipment be stolen, or a site evacuated, it may be desirable to disable the VSAT modem temporarily by stunning or permanently killing it. Only the network operator or factory can reactivate a system so disabled. This prevents the system being used by unauthorized or undesirable users and is an option in some modems.

19) <u>Possible equipment choices.</u> Remote modem equipment falls into several levels largely determined by the target market. The delineations are not solid, with crossovers of some products into adjacent markets.

   a) The consumer market is aimed at IP connectivity to users with one or two personal computers (PCs) on the LAN. Their connectivity is seldom mission-critical and they can accept slowdowns and stoppages on the network. To deliver a very low monthly cost, service providers massively oversubscribe the space segment, to as high as 80:1. Terminal cost is also very low. They are classed as "best effort" systems and have no SLA guarantees. Typical platforms include Hughes DirecWay, Gilat 360E, and ViaSat Surfbeam.

   b) The small office home office (SOHO) market requires more reliable connectivity with less slowdown and no stoppage of data. These networks typically support 2-10 PCs on the LAN, and also rely largely on a shared bandwidth model. These systems may support a single VOIP phone line as well, with minimal packet loss or jitter. Over-subscription is much lower, typically in the 10:1 to 30:1 range. They may or may not be a true best effort system, and many of the lower cost services have no SLA. CIR service is sometimes an option, but is outside the normal network. Common platforms include the ViaSat Linkstar and Gilat Skyedge.

   c) The Small-Medium Enterprise (SME) market is for commercial users with larger networks (10-100 PCs) on the LAN, or who require faster throughput, or both. They also can support 1-8 VOIP lines based on satellite bandwidth with very high call quality. These networks almost always have an SLA in place with remedies enforced for under performance. The industry standard over-subscription for shared SME service is 8:1. The network also can incorporate sites with CIR and SCPC. Typical hardware is the iDirect iNFINITI Series 3000 and 5000 modems.

   d) The large corporate user market relies heavily on high-bandwidth SCPC CIR services. They can support 50-500 PCs and 20-100 VOIP calls. Remote terminals include the Gilat commercial series, Comtech/EFData products, ND Satcom, and the iDirect Series 7000 modems.

VoIP Considerations

VoIP capability is increasingly in demand. The satellite hardware platform chosen should be able to support from one to four simultaneous calls, depending on the need at the remote sites.

1) Minimize latency. Choose a network and equipment that adds minimum delay. The latency of the satellite link adds fixed delay of typically 560 mSec. Additional fixed (but typically minimal) delay is added by the satellite transponder. The VSAT hub and client remote equipment causes more delay. The satellite network topology and equipment should be chosen to add minimal additional latency to the end path.

2) Avoid jitter. VoIP relies on a constant stream of data being transmitted sequentially. Packets transmitted at equal intervals may arrive at the destination with irregular delay and out of sequence. This is called jitter, and it must be managed and controlled to ensure good VoIP voice quality. This can be accomplished by sending the data in short transport frames and distributing them evenly in the overall data stream, by dynamically allocating bandwidth several times per second, by fragmentation so that large data packets don't get in front of small voice packets, and by managing the queue depth. Choose a modem platform and service that transfers VoIP calls under CIR dedicated bandwidth.

3) Minimize packet loss. VoIP traffic (UDP/IP) is not retransmitted like TCP/IP traffic, so it is lost if an error occurs. Very low Bit Error Rates (BERs) are required for high-quality voice calls.

4) Seek bi-directional QoS and traffic prioritization. Satellite networks are subject to congestion, especially those using shared bandwidth. This can lead to delayed, dropped, or out-of-sequence packets. Bi-directional QoS ensures delivery of VOIP traffic on congested networks.

5) Choose adequate compression. The dominant industry standard for VoIP calls is the G.729 CODEC, requiring a minimum of 8 kbps of bandwidth. When IP/UDP/RTP headers are added, the actual bandwidth required is 16-18 kbps. Modems that compress the RTP drop the bandwidth requirement to 10 kbps.

6) Have enough bandwidth for VOIP calls. G-729 CODECs require 16-18 kbps per call uncompressed, and 12 kbps if compressed. G.723 will require about 10-12 kbps once compressed. Incidentally, video-conferencing at 30 frames per second requires about 384 kbps of CIR, in both directions, for excellent quality.

7) Internal corporate networks. Corporate users may prefer to run the VoIP link directly to their main offices. A gateway at that location will tie the remote site into the PBX at the main office. The remote-site telephones will function as extensions on the main office PBX. This can be accomplished through either a small PBX at the remote site, or with direct single-line phones. Calls into and out of the internal network and the PSTN are handled like any other calls on the phone system.

8) Using a public gateway. VoIP calls can be routed to a public service for interfacing to the PSTN. The gateway services can be run by private companies or may be part of a service offered by the satellite service provider. Calls can be made into and out of the PSTN and satellite remote sites.

9) Billing options. If an external service provider is used for VoIP, it should offer a number of options for billing, including post-paid billing, pre-paid accounts, and pre-paid scratch cards.

Considerations in Purchasing a VSAT System

Purchasing a VSAT system is not the same as buying a DSL or cable modem connection at home. Nor is the performance comparable. A 512 kbps VSAT connection should not be compared with a 512 kbps DSL or cable modem connection. To purchase a satellite system that will perform cost-effectively, it is important to understand user needs, the operational environment, and the cost-performance trade-offs. Some factors to be considered are outlined below.

Geographic Area of Operation

The geographic location of the operation will determine the performance and services that are available from the commercial satellites that cover each part of the world. The available performance and services will in turn determine:

- Viability: If the satellite to be used is too low on the horizon—or obscured by mountains, tall buildings, or trees—it may be necessary to use an alternative.
- Suitability: In areas of high rainfall, such as the tropics, C-band systems will provide more consistent service than Ku-band systems during rainfall. But C-band service typically requires a bigger, more expensive dish. The signal strength provided by different satellites at any location may also point toward one option more than others. The power rating of the electronics that are fitted to the satellite dish should be modified to match the available signal.

Satellite Connection and Use Considerations

The characteristics of Internet traffic created by different applications are different, and they give rise to differing performance specifications for the connection to be purchased. For example, VoIP (and particularly, video conferencing) is a demanding application. It calls for properly specified connections to work well. Even Web surfing and email become painfully slow unless the connection performance is adequate.

The number of people that will typically use the connection simultaneously needs to be determined. For example, if there is a 512 kbps connection at 50:1 contention, each user might get as little as 10 kbps at peak times. If each of these subscribers has 10 users on its network, each user might then get as little as 1 kbps of bandwidth. Another performance factor is upload speeds, which are usually lower than the download speeds. For example, if the upload speed is nominally 128 kbps, each user might get as little as 0.1 kbps in upload capacity. Sending a 400 kilobit file under these circumstances would be operationally unacceptable.

Toleration of Performance Variability

If dedicated bandwidth is purchased, with a guaranteed throughput that is CIR service, the full performance will most likely be available when it is needed, but it will be very expensive. It can be much more affordable to purchase a highly contended link with a "best effort" contract, but users will probably be very frustrated with the lack of performance. The challenge is to balance the cost against the system's ability to meet performance requirements.

Another major consideration is quality of service. QoS management enables different applications to be given different priorities for capacity utilization. With applications like VoIP, it is critical that data be received promptly and in the right order. Otherwise, conversation will be unintelligible. With other applications such as email, the need for impeccable reception of packets is less critical. QoS management would give VoIP data priority over email data, increasing the usability of the VSAT link. Other factors can provide unanticipated interruptions to service:

- "Rain fade" – Rain can interrupt Ku-band service (especially in the tropics), while C-band transmission is unaffected.
- Duration of service – Interruptions due to equipment failures can require on-site technical support, which will depend on contractual terms for logistics support and supplier response.

Some satellite terminal equipment is pretty basic, and it may be necessary to buy equipment in addition to the satellite modem to make the connection work adequately. Other manufacturers may have a range of additional functionalities built in, saving money. So be sure to compare like capabilities when making selection decisions.

Select a system configuration that balances performance expectations and cost constraints. Considerable economies of scale can be realized by purchasing in large volumes and by committing to long-term contracts. Be careful not to lock yourself into high rates, however, because bandwidth costs are on a downward trend. Limit the bandwidth passing through the connection by the use of features such as:

- Proxy caching;
- Spam and content filtering at the hub (before it goes over the satellite link);
- Web compression;
- Bandwidth efficiency/throughput enhancement technologies built into some equipment.

Best Practices for Conducting Operations

This section provides best practices and guidance for the period during which a deployment takes place. There is an overall need for a common operational picture and enhanced situational awareness. This calls for the creation of a common, relevant, releasable operational picture that can be used by all responders to an HADR or S&R scenario. Accordingly, information should be collected, organized, and disseminated in a manner that will benefit the population and polity of the affected host country, within the operational goals.

ICT Infrastructure Assessments

• <u>Assessments of ICT infrastructure damage should be placed in context and related to pre-existing or baseline data to determine the actual extent of destruction from the conflict or disaster event.</u>
Assessments of baseline ICT capabilities before a conflict or disaster should consider several aspects, including:

- o Telecommunication policies, regulations, and laws;
- o Telecommunication procurement rules and business practices;
- o Infrastructure standards, capabilities, coverage, connectivity, manufacturers, and supply lines;
- o Telecommunications business practices;
- o The technical and management skill base; and
- o The information culture of the host nation, including how information is gathered, shared, and used.

Assessments of the state of ICT infrastructure after the conflict or disaster must consider:

- o What are the rule-of-law, government, and population communications needs that must be met?
- o What are the communications media that are currently available, functioning, or in need of repair?
- o Can the host nation ICT infrastructure be updated to employ modernized technology?
- o What communications assets and actions can the military provide to stabilize and reconstruct the infrastructure?
- o Are telephone systems configured to modern standards?
- o How many main telephone lines are in use?
- o Are mobile and satellite communications operational?
- o Does the population have access to Internet capabilities and providers?
- o What is the ability of the ICT infrastructure to accommodate responders' surge demands?

• <u>Assessments that are strictly qualitative (narrative) tend to lose value over time and are difficult to integrate with other forms of data.</u>
Consider, for example, the following statement:

> "The assessment team reports a need to prioritize reconstruction of the two destroyed bridges that connect agricultural producers with the local market. A third destroyed bridge prevents access to the nearest port."

To provide added value, assessments should include location information in the form of coordinates. The previous statement, with coordinates added, would look like this:

> "The assessment team reports a need to prioritize reconstruction of the two

destroyed bridges (2.5700399, 96.0752737), (2.5838459, 96.0315965) that connect agricultural producers with the local market. A third destroyed bridge (2.5625190, 96.0862688) prevents access to the nearest port."

Then, a database entry, similar to table 2, can be developed, using the coordinates to include additional reconstruction information:

RECNUM	LAT	LON	LONGNAME	Type	Partner	Start_date
9	+2.5700399	+96.0752737	1	concrete piers	Atlas Bridges	5/5/2005
36	+2.5838459	+96.0315965	2	steel span	Bridge Tek	4/4/2005
49	+2.5625190	+96.0862688	3	concrete piers	Atlas Bridges	6/6/2005

Table 2, Sample Database Entry

In addition, a map like the one shown in figure 14 can be produced from the database coordinates:

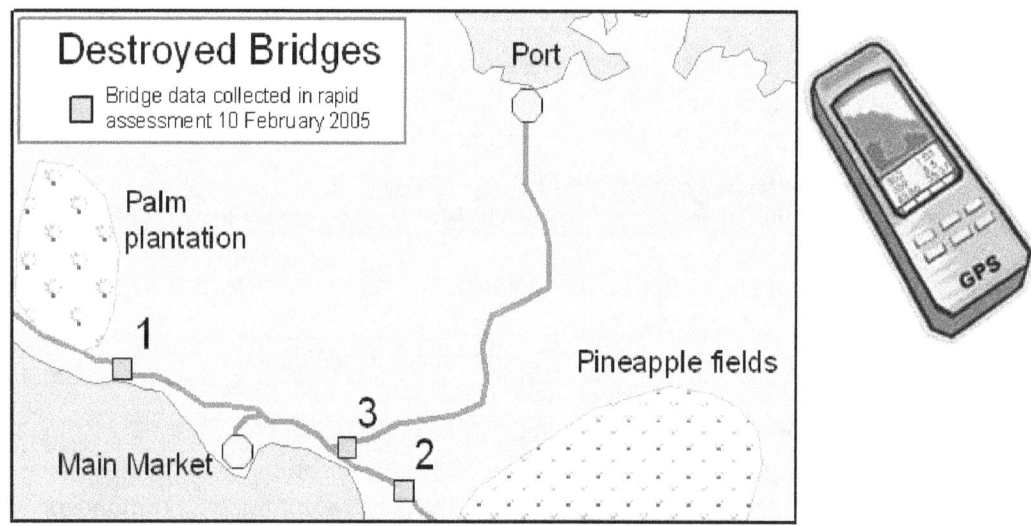

Figure 14

Geo-referenced (location) data is collected using a GPS unit. GPS units are now easier to use than ever before. Some require only the pushing of a button to capture location coordinates. These coordinates can be downloaded directly from the GPS unit into a computer and used to produce maps and spreadsheets or databases.

Collecting location data increases both the immediate and long-term value of the assessment. Having the location information reduces the need for multiple visits to the same area to determine exact locations. In addition, with location information as a common reference, additional data sets (agriculture, roads, and markets) can be added to

94

the map. Geo-referenced information then becomes useful to many actors in all stages of a complex humanitarian emergency.

Field Coordination

Planners should encourage broad ICT development and participation from local, national, and international civilian and military actors to facilitate and support S&R and HADR response activities.

• <u>All parties should foster partnerships with specialized agencies and sector experts to conduct sectoral surveys and analyses before events occur.</u>
Information-related efforts that are incrementally resourced and initiated only as emergency situations unfold tend to remain behind the curve and reactive. This leads to a failure to provide timely information that is accurate and contextual.

• <u>Preparedness measures, such as base data preparation for high-risk areas, national-level capacity building, and the formation of institutional relationships should take place prior to deployment.</u>
This will enable information management and exchange systems to effectively support assistance efforts once an emergency begins. Preparation also includes planning for sustainability and/or exit strategies.

• <u>Planners from all participants should pre-establish agreements for data-sharing.</u>
Participants must collect and maintain their own baseline data, but they should agree upon a central repository for all collected information. Participants can employ, where possible, standardized formats. A common metadata catalogue, providing standards for, and documentation of, metadata (who collected what information where, and when), should be established. Facilitation of data management should be geared to add value to existing information, rather than replacing or duplicating it.

• <u>Information repositories should be configured so that all organizations have equal access to the information, which can be facilitated by using standard formats.</u>
Such repositories can be viewed as conceptual toolboxes that collect tools, principles, and best practices. Moreover, it is essential to define public domain information at the outset, clearly differentiating public and private information. Archives and archiving procedures will serve better if they are established early, eliminating confusion. Information management systems should meet the clearly defined needs of operational users and decisionmakers and aim to reduce the effects of information overload. Data and information must be relevant, accurate, and timely.

• <u>Ensuring a quality product requires the development of, and adherence to, standards for information collection, exchange, security, attribution, and use.</u>
It is vital to maintain a strong sense of professional ethics at every stage of information system design and implementation, including such elements as independence and impartiality, in pursuit of humanitarian action. Multiple information systems, including Web sites and databases, operating at global, regional, and local levels, create the

potential for an unprecedented degree of cooperation between organizations and people at the field level, between the field and headquarters, and between the international and local communities. Partnering with the media can be an effective way of communicating information to the affected population.

A Field Data Collection Hierarchy

This subsection offers a default standard or format for data collection. The order of data collection is important. For example, data in items 1 and 2 of table 3 are imperative for use in GIS analysis and mapping. Additional items are essential for the validity and relevance of the data; and for enabling it to be shared. Table 3 lists several of the standard items:

1) Where: Location information and how it was collected
   a. Latitude
   b. Longitude
   c. Feature
   d. Location
   e. Collection Accuracy
2) When: When the data was collected
   a. Date
   b. Time
3) What: Details of what was happening at that particular location
   a. Sector
   b. Project Name/Title
   c. Comments
4) Who: Who collected the data
   a. Agency/Organization
   b. Collector (individual)

Data Table Items: (*individual definitions provided below table*)

Latitude	Horizontal Position (x-axis)
Longitude	Vertical Position (y-axis)
Feature	Short description of item being assessed
Location	Most precise name possible
Collection Accuracy	How the location is determined
Date	Day/Month/Year
Sector	Select from pre-defined list
Project Name/Title	Project is referred to as
Comments	Supplemental information
Agency/Organization	Funding Source
Collector	Person or Team collecting data

Table 3

The following explanations and best practices are provided, regarding each of the data fields in table 3:

Latitude and Longitude: Data collected for latitude and longitude provide geo-spatial coordinates for a specific point. An example is collecting a point to represent the location of a hospital. This data is used to create a *Point GIS layer*. A *Start* point and *End* point location is collected for linear projects so that a *Line GIS layer* can be created. For projects that encompass a region, the center point (centroid) of that region should be entered in latitude and longitude, and the description of the region should be added into the Feature field. Latitude and longitude are collected using decimal degrees in a WGS 84 projection, and they can readily be collected using a GPS device. These devices can be purchased for approximately $150. They include software tutorials for learning how to use them.

• It is strongly recommended that all data collectors purchase and learn how to use a GPS device.

Feature: A description of what the latitude and longitude refers to. Examples include: Health Center, Food Distribution Site, Village, School, Hospital, Bridge, Port, Highway, and other such designations.

Location: To facilitate mapping, use the most precise name possible in order of preference: town/village, city, district, administrative region, state.

Collection Accuracy: This describes how the positional information was gathered. Items in this field include details of accuracy levels and sources of the positional location. Examples: GPS – 30m accuracy, USGS topographic 1:50000, Military Grid Conversion.

Date: This is the date the data was collected. The date field should be in the format: day, month, year.

Sector: This refers to functional field classification of projects into defined categories: health, transportation, education, justice, and others. Definitions are supplied by the survey agency to meet their reporting requirements. Data collectors should be provided with a pre-defined list of sectors to use before filling out data.

Project Name/Title: Official title or name of the project.

Comments: Additional information as needed. Example: "This school building is used in three shifts for three different schools. Therefore the same Latitude & Longitude refers to Salem School for Girls, Boys Secondary Training, and Bangle Nursing School." This field is not to be used in lieu of filling out the required fields.

Agency: This refers to the agency or organization funding the project, for example USAID, UNHCR, or others as applicable.

Collector: Name of the agency, unit, organization, or contractor responsible for collecting the data entered in this table. This is important for verification and clarification.

Employing a Website

Responders (both military and non-military) often start a website or add a page to their existing website to spotlight their activities in a crisis area of operations. Organizations find at least two reasons to do this. First, a website provides a virtual location to exchange information and post data so that it is accessible to parties both in the crisis area and at the organization's headquarters. Second, the website may provide highly sought public exposure to the organization's humanitarian aid work, resulting in greater political clout and fundraising ability. The following general best practices are offered with regard to a website outreach strategy:

- Serve the needs of the target audience, including international and local organizations;
- Remain focused and simple;
- Link to existing systems to avoid duplication of effort;
- Indicate the source of the information and include disclaimers to guard against inaccuracies in data collected by other parties;
- Keep up with technology, but never give technology priority over content;
- Preserve institutional memory (update, retool, archive);
- Facilitate humanitarian coordination;
- Improve operations;
- Inform decisionmaking;
- Adhere to principles of humanity, impartiality, neutrality, and independence;
- Base the content on needs of practitioners and decisionmakers and include website elements such as:
  - A "resource center" that includes a document library, GIS map center, imagery library, pictures, white papers, field reports, and database;
  - A "community area," including a notice board, contact lists for all organizations involved;
  - A site map;
  - A section listing vacancies for operational positions;
  - A list of relief guidelines, emergency procedures, local regulations and procedures that may affect operations;
  - A listing of support services and contacts;
  - Weather and situational awareness (security and status of infrastructure) information;
  - News (local and international) relevant to crisis response actions; and
  - Links to relevant and credible news outlets, blogs, other websites, urgent appeals, aid offers, announcements, and other information sources.
- Avoid using heavy graphics, including embedded pictures, in files to be downloaded;
- File sizes should be indicated so that users can decide whether or not to take the time needed to download information;

- Summarize the content of files and provide user comments on the usefulness of the information;
- Provide an ability to assess, filter, and tag most useful information;
- Follow a pre-established set of standards that guide:
  - Procedures, including information collection methodologies and language;
  - Technology, including the use of hardware and software packages to ensure maximum interoperability and development and maintenance by local staff;
  - Data, including standards for ensuring data formats, content, and quality;
  - Metadata, including pre-established and common standards for identifying and documenting data.
- Involve partners and stakeholders in the early conceptualization and development of the website;
- Promote the website and "sell" the benefits of its products and services on a regular bases once the site has been established;
- Education and training enhances buy-in; and
- Key institutions and partners should develop common policies governing the establishment and maintenance of websites.

Planning and Building Wireless Networks

The publication *Wireless Networking in the Developing World: A practical guide to planning and building low-cost telecommunications networks* is a best practice on how to build an ICT infrastructure using commercial technology to serve as the backbone for wide area wireless networking. By applying commercially available ICT in areas that are badly in need of critical communications infrastructure, more people can be brought online than ever before, in less time, and for very little cost to support voice communications, Internet access and email and other data exchange. The book illustrates how to make such networks work, and provides information and tools needed.

Wireless Networking in the Developing World can be freely downloaded as PDF files as a whole (9.4 MB) or in single chapters (ranging from 72 KB to 976 KB) from the web site (http://wndw.net/). The book table of contents shows these sections:

About this Book
Where to Begin
A Practical Introduction to Radio Physics
Network Design
Antennas & Transmission Lines
Networking Hardware
Security
Building an Outdoor Node
Troubleshooting
Case Studies
Appendices

The book and PDF file are published under a Creative Commons Attribution-ShareAlike 2.5 license. This allows anyone to make copies, and even sell them for a profit, as long as proper attribution is given to the authors and any derivative works are made available under the same terms. Any copies or derivative works *must* include a prominent link to their website, (http://wndw.net/).

See (http://creativecommons.org/licenses/by-sa/2.5/) for more information about these terms. Printed copies may be ordered from Lulu.com, a print-on-demand service. Consult the website (http://wndw.net/) for details on ordering a printed copy. The PDF will be updated periodically, and ordering from the print-on-demand service ensures that you will always receive the latest revision.

Trends in the Use of Commercial ICT

In spite of nearly global coverage of commercial ICT networks depicted in figure 15, commercial ICT capacity specifically dedicated to serving the needs of many of the civil-military responders to disasters and recovery efforts has been operating mostly in a vacuum. Additionally, the military has been reluctant to rely on commercial ICT capacity for C2, although commercial ICT has been used for combat support missions such as administration, logistics, and medical tasks. The operational environment is changing, however, and commercial ICT capacity is becoming more important, not only for civilian responders but also for the military. Civilian and military elements are both relying on commercial ICT flyaway packages to support forward-deployed responder elements. Commercial products also play an important role in supporting ICT capacity for long-term military operations and civil-military reconstruction and development of the affected nation's ICT infrastructure.

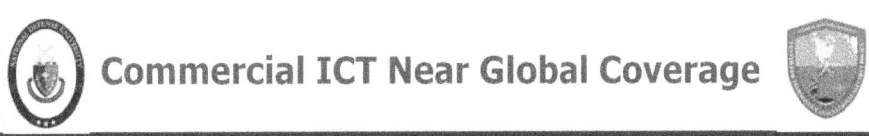

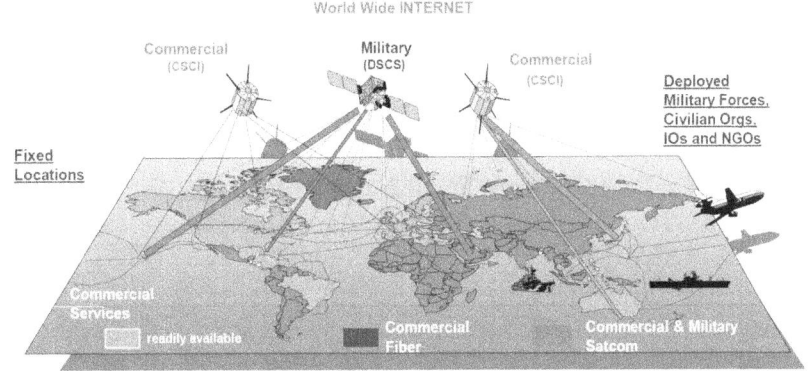

Figure 15

Traditionally, the military employ "ruggedized," purpose-built tactical systems to support C2 and to extend ICTs forward into a crisis area. A dedicated fixed and mobile ICT military infrastructure was used to create a secure CIE to support combat operations, early interventions in HADR operations, and related support missions. But these systems have not been designed to fully support long-duration operations related to affected nation recovery, reconstruction, and development or to provide services to non-military elements involved in the disaster response effort. Unclassified military networks are connected to the Internet, and these links can be used to informally exchange information with non-military participants in a crisis without connecting these participants directly to the military networks. The military can maintain a secured firewall interface with Internet.

Use of .mil email addresses and website URLs does, however, create a problem for many NGOs that are reluctant, especially in hostile operational environments, to be seen as directly exchanging information with the military. It would be better for the military to use a neutral surrogate means, such as hotmail or yahoo email accounts and third-party impartial web sites such as UN OCHA ReliefWeb or some other neutral (non-military) web site to post and exchange humanitarian information.

The military has started to employ some commercial mobile ICT capability packages to support limited remote access to the Internet, for example, the Joint Task Force (JTF) Commander Executive C2 package illustrated in figure 16. To accommodate a more neutral means of exchanging information with NGOs, the military may need to consider alternative means, for example, a separate and independent network for interfacing with non-military participants. For longer duration operations, the U.S. military does commercialize its deployed fixed-base ICT support arrangements to replace the tactical military systems so that the tactical systems can be made available for future military operations. Here too, it may be necessary to consider alternative means for communicating and sharing information with non-military elements.

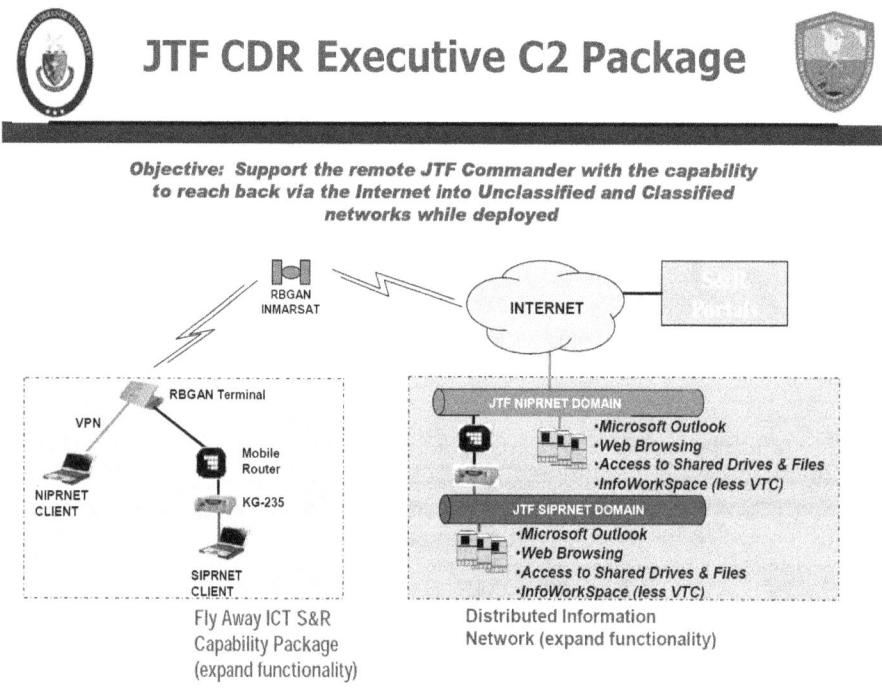

Figure 16

The non-military actors—civilian government agencies, IOs and NGOs—rely largely on commercial ICT products and services to extend communications and information services to a crisis area. Some have their own limited ability to establish ICT connectivity forward. These are mainly packages consisting of a mix of commercial HF/VHF/UHF radios, GSM cell phones, satellite phones and laptops, sometimes with WiFi capabilities. Organizations such as the UN manage their own global IP networks to support headquarters organizations, and the deployed UN agencies use commercial ICT packages to connect to these networks from crisis areas. The forward deployed ICT capabilities can be provided by the UN FITTEST for agencies like WHO and UNJLC or, as is the case for UNDPKO, they can be provided as part of the organization's indigenous ICT capabilities as shown in figure 17.

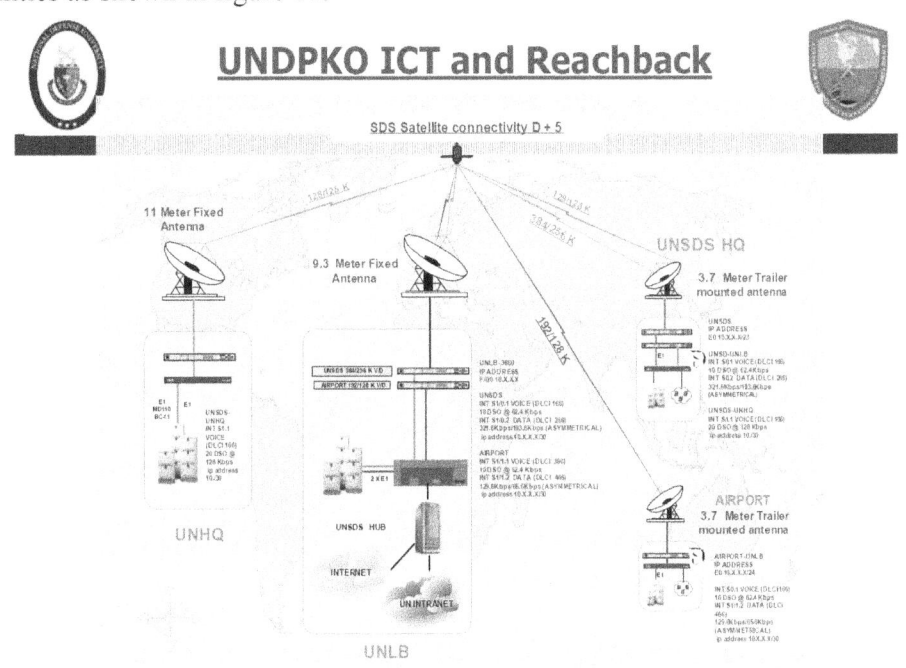

Figure 17

Other non-military participants purchase or lease portable and fixed ICT capabilities and employ them to access the Internet via commercial satellite services and to support local wired and wireless access arrangements. These capabilities also are used to set up Internet cafes that provide service to other responders, as well as to the local population and leaders. Companies, such as I-Linx (www.i-linx.net), ProActive (http://204.200.201.151/) DRASTIC (www.drasticom.org), and others can provide a wide range of portable and fixed installation arrangements. Such arrangements can provide: satellite access, for example on an RBGAN or VSAT terminal; WiFi routers; portable solar panels for power; laptops; PDAs; Internet access; and other field needs such as tents, tables, chairs, and technical support. Longer-term fixed networks to support e-commerce, e-government, and other IT services can be provided as turnkey or managed services.

Non-profit organizations such as HumaniNet (www.humaninet.org) and Network Startup Resource Center (www.nsrc.org/) offer humanitarian relief and development field teams that give NGOs practical assistance and advice on ICT implementation and operations. The Global Disaster Infrastructure Networks (www.gdin.org) is a non-profit web site used to share disaster information with interested participants with an aim to providing the right information, in the right format, to the right people, in time to make the right decisions.

The Internet and related portals are used to create a distributed information environment that can support coordination, assessments, and information sharing. Some portals, such as those managed by the UN, provide information on global disasters and relief activities. Some examples of these portals are the UN OCHA Integrated Regional Information Networks (www.irinnews.org) and its ReliefWeb (www.reliefweb.int); the joint UN OCHA and European Commission Joint Research Center (http://www.jrc.cec.eu.int/)and Global Disaster Alert and Coordination System (www.gdacs.org); and the UN WFP (www.wfp.org). The U.S. Pacific Disaster Center (www.pdc.org/iweb/index.jsp) and U.S. Geodetic Survey (http://www.ngs.noaa.gov/) are national examples of a similar kind of portal.

Other portals are developed to respond to specific disaster situations, for example the UN HIC (www.humanitarianinfo.org), the DOS eRoom (http://hiu.state.gov), and the UN Afghanistan Information Management System (www.aims.org.af).

Portals are also being used as "virtual operations centers." These include the UN OCHA Virtual On-Site Operations Coordination Center (http://ocha.unog.ch/virtualosocc/) and the U.S. Pacific Command Virtual CMOC and Groove-based Virtual Emergency Operations Center set up in response to the Tsunami.

NGOs are also developing portals. The Vietnam Veterans of America Foundation (http://www.vvaf.org/) has a prototype portal being used in Iraq to share security situation awareness information with civil-military participants. It is being developed for future use in other crisis operations as well. Bloggers also have surfaced as a cottage industry providing observations, information, and assessments on global disasters.

Both synchronous and asynchronous collaboration tools are emerging. The U.S. military is using InfoWorkSpace, a synchronous collaboration tool, and Groove, an asynchronous collaboration tool that is also used by some civilian organizations. Other software tools, such as NetMeeting, WebEx, and POPG, are also used. None of the current tool suites is tailored to meet the aggregate needs of the civil-military participants, and the current tool suites are not interoperable. The challenge is to develop an interoperable tool suite that will accommodate both synchronous and asynchronous modes of operation.

VoIP is fast becoming a popular means for providing voice services in austere environments, both for communicating with other responders and with headquarters operations. Skype is one such capability. VoIP also can be used to set up local communications where wireless data networks are deployed. Such communications can be used to serve local populations, as well, through the use of call centers or by providing local users direct access.

Language is a major challenge in the operational environment. Several tools are emerging to address automated language translation. Examples of such software tools are Virage (www.virage.com/home/), developed by Virage Inc., Babylon, Translation Information Detection Extraction Summarization (TIDES), developed by DARPA.

Recent natural and manmade disasters have given industry and NGOs greater awareness of business opportunities in the field of providing mobile and fixed ICT capacity for deployed responders. The same is true for supporting post-disaster reconstruction and development of the affected nation's long-term ICT capacity. IT and telecommunications-oriented NGOs and private businesses are now offering lightweight, rapidly deployable ICT packages for victims and aid workers. These packages often are available to anyone in the crisis area: military personnel, governmental agencies, IOs, NGOs, and local citizens and community leaders. Other companies and NGOs are addressing the market for longer-term ICT support of reconstruction and redevelopment. These quick-reaction and redevelopment support packages increasingly can be acquired through separate equipment purchases or as a service from an ICT service provider or system integrator. Figure 18 illustrates some examples of the range of capabilities being offered:

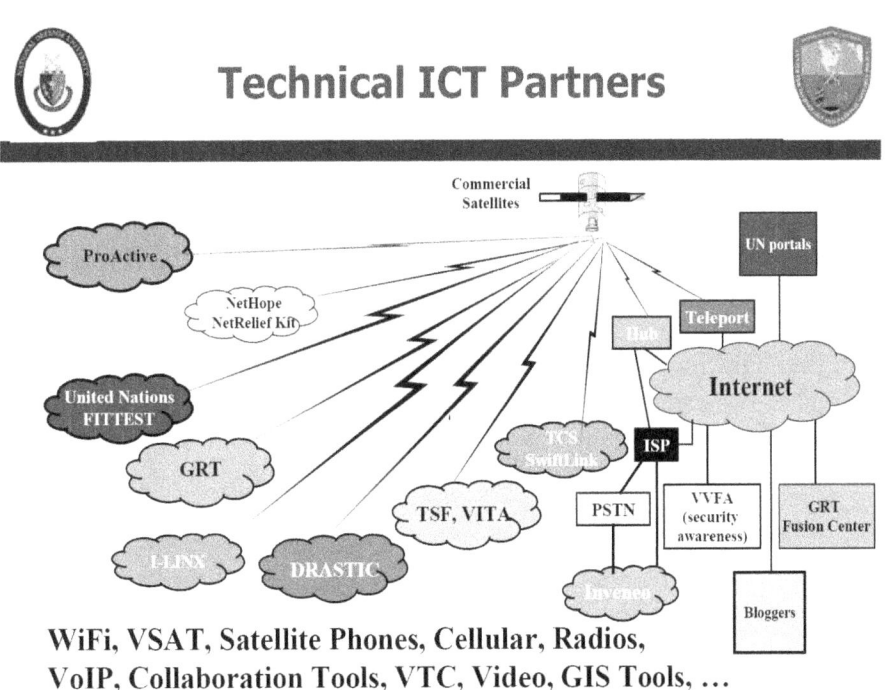

Figure 18

NGOs such as NetHope (http://www.nethope.dreamhosters.com/index.html)and their "Net Relief Kit" provide a type of Internet-in-a-box capability that is easy to set up and operate as a communications hub for disaster management. These products can support voice service, Internet access via a commercial satellite connection, and local wired and wireless access. Other NGOs, such as Télécoms Sans Frontières (TSF), offer telecommunications facilities anywhere terrestrial infrastructure is destroyed, insufficient, or non-existent. The organization Volunteers in Technical Assistance (VITA) provides low-cost communications and information services through a commercial satellite connection.

TeleCommunication Systems' "Swiftlink" (www.telecomsys.com) is an example of a rapidly deployable communications package that provides a secure reach-back capability to the U.S. military SIRPNET and NIPRNET networks through a commercial satellite connection. The product also can be used by non-military elements for communications and Internet access.

Inveneo (http://www.inveneo.org/) offers a solar-powered wireless network capability (referred to as a "solar village") that supports voice and data services for NGOs and local populations in severely limited ICT environments, such as those without local power sources.

Other industry providers, such as Global Relief Technologies (GRT) (www.globalrelieftech.com), have targeted a particular niche area like remote data collection, fusion, and information dissemination. GRT provides a field data collection and situational awareness service that employs PDAs with preformatted templates that can be used to collect data in internationally agreed formats with geo-referencing (and free text and digital pictures). The data is then up-linked through a commercial satellite connection to a "fusion center" in the United States, where it is aggregated, assessed, and displayed on maps. The information stored at the fusion center is Web-accessible, so authorized people in the field and elsewhere can gain situational awareness information via the Internet. The fusion center operates 24 hours a day, seven days a week.

The examples cited in this section are by no means all-inclusive; they represent a small sampling of the NGO and industry offerings of ICT products and services. Other companies and NGOs can provide similar ranges of services—indeed, this is an ICT industry market growth opportunity area, and new companies and service and product offerings continue to proliferate.

Conclusion

The consensus is nearly universal that ICTs are necessary enablers for the smooth, effective operation of humanitarian relief and S&R missions. Moreover, a growing number of participants in these operations, from military units to the smallest NGOs, are bringing at least some elements of their ICT capabilities with them when they deploy. Yet the watershed events of recent years—the Indian Ocean tsunami, the rash of Caribbean hurricanes, and the devastating earthquakes in Iran and Pakistan—show that there is no default or standardized suite of equipment, databases, or operational protocols to follow when diverse organizations attempt to work together for humanitarian purposes.

This primer has explored the increasing amount of commercial ICT technology now available to relief and reconstruction workers. Telecommunications and information management systems can be engineered to be complementary and interoperable. But this primer also has explored the organizational and institutional gaps that persist, particularly across the civilian-military boundary. No easy answers will be found to the operational problems that can result from disconnects among the various military and civilian actors. But common ground can be reached, built around the precept of unity of effort in humanitarian disasters, post-conflict S&R scenarios, and complex emergencies.

In other words, technology has a human element beyond the wiring and software packages that are commonly regarded as comprising it. The very nature of natural and man-made disasters means that many individuals and organizations will have a stake in responding to them and ameliorating their effects. No single, unified command structure is likely to fit every situation. An interlocking set of command structures, representing multiple groups, nationalities, and interests will remain. Establishing a CIE in such scenarios is a challenge to the very core of what information and communications technologies are supposed to be in the 21st century: distributed, decentralized, flexible, and interoperable. From the perspective of the military, establishing a CIE in HADR and S&R operations is a key test of the entire concept of net-centricity—pushing power and intelligence to the edge of networks. For the NGOs and IGOs, the use of ICTs offers what they often desperately need: a multiplier effect from the application of expertise and limited resources.

This primer has been born in the spirit of matching the twin drives of net-centricity and the enabling of humanitarian aid workers in the interest of improving the efficiency and effectiveness of HADR, S&R, and complex emergency operations. To the extent that responders will always need to coordinate, this primer does not come to an end. In truth, this work is only a beginning.

Appendix A: Glossary

Buffer zone:
DOD - A defined area controlled by a peace operations force from which disputing or belligerent forces have been excluded. A buffer zone is formed to create an area of separation between disputing or belligerent forces and reduce the risk of renewed conflict. Also called area of separation in some UN operations.

NATO - A designated area between the positions of the parties in conflict, agreed upon by the states concerned. The buffer zone is restricted exclusively to individuals, agencies, forces, and bodies specially authorized by the organizers of the zone.

UK - The neutral space between ceasefire lines.

UN - Also known as area of separation; neutral space created by withdrawal of both hostile parties; a demilitarized zone where the parties have agreed not to deploy military forces; the cease-fire lines, marked and often fenced or wired on either side of the buffer zone, indicate the agreed forward limits of the contending forces; the cease-fire lines are observed, patrolled and perhaps occupied by the peace-keeping force; the buffer zone itself may be placed under the control of a PKO.

Civil affairs (CA):
DOD - 1. The activities of a commander that establish, maintain, influence, or exploit relations between military forces and civil authorities, both governmental and non-governmental, and the civilian population in a friendly, neutral, or hostile area of operations in order to facilitate military operations and consolidate operational objectives. Civil affairs may include performance by military forces of activities and functions normally the responsibility of local government. These activities may occur prior to, during, or subsequent to other military actions. They may also occur, if directed, in the absence of military operations.

DOD - 2. Designated Active and Reserve component forces and units organized, trained, and equipped specifically to conduct CA activities and to support CMO.

Civil-military operations center (CMOC):
DOD - An ad hoc organization, normally established by the geographic combatant commander or subordinate joint force commander, to assist in the coordination of activities of engaged military forces, and other USG agencies, NGOs, and regional and international organizations. There is no established structure, and its size and composition are situation dependent. See also civil affairs activities; civil-military operations; operation.

UN - In a PKO which contains substantial civilian elements, a civilian-military structure of integrated support services may be established to perform liaison and coordination between the military support structure, NGOs, PVOs, and local authorities; for example the CMOC was opened by U.S. JTF Support Hope in Entebbe (Uganda) and Kigali (Rwanda); in Haiti, UNMIH has an integrated civilian/military headquarters.

Civil-military cooperation:

DOD - All actions and measures undertaken between NATO commanders and national authorities, military or civil, in peace or war, which concern the relationship between allied armed forces and the government, civil population, or agencies in the area where such forces are stationed, supported, or employed.

NATO - The resources and arrangements which support the relationship between commanders and the national authorities, civil and military, and civil populations in an area where military forces are or plan to be employed. Such arrangements include cooperation with non-governmental or international agencies, organizations, and authorities.

Civil-military operations (CMO):

DOD - 1. The activities of a commander that establish, maintain, influence, or exploit relations between military forces, governmental and nongovernmental civilian organizations and authorities, and the civilian populace in a friendly, neutral, or hostile operational area in order to facilitate military operations and to consolidate and achieve operational U.S. objectives. CMO may include performance by military forces of activities and functions normally the responsibility of the local, regional, or national government. These activities may occur prior to, during, or subsequent to other military actions. They may also occur, if directed, in the absence of other military operations. CMO may be performed by designated CA, by other military forces, or by a combination of CA and other forces.

DOD - 2. Groups of planned activities in support of military operations that enhance the relationship between the military force and civilian authorities and population and which promote the development of favorable emotions, attitudes, and behavior in neutral, friendly, or hostile groups.

USAID - Activities in support of operations including the participation of both the military forces and civilian authorities.

Combined joint task force (CJTF):

NATO - A multinational and multi-service force, task-organized and formed for contingency operations, which requires multinational and multi-service command and control. Designed to make NATO's joint military assets available for wider operations by NATO nations or by the Western European Union (WEU), the concept of CJTF was introduced at the 1994 Brussels Summit. Detailed work continues in coordination with the WEU on the implementation of the concept, with a view to providing separable but not separate military capabilities that could be employed by NATO or in operations with nations outside the Alliance.

UN - Concept approved by NATO leaders in January 1994, under which U.S. materiel and forces designated for NATO operations can now be made available for non-NATO activities in out-of-(NATO) area operations, such as those humanitarian relief or peacekeeping operations initiated by WEU to deal with regional instabilities and ethnic conflicts.

Combined joint special operations task force:
DOD - A task force composed of special operations units from one or more foreign countries and more than one U.S. Military Department formed to carry out a specific special operation or prosecute special operations in support of a theater campaign or other operations. The combined joint special operations task force may have conventional non-special operations units assigned or attached to support the conduct of specific missions.

Communications and information system:
NATO - Assembly of equipment, methods, and procedures, and if necessary personnel, organized so as to accomplish specific information conveyance and processing functions.

Communications network:
DOD - An organization of stations capable of intercommunications, but not necessarily on the same channel.

Communications zone:
DOD - Rear part of a theater of war or theater of operations (behind but contiguous to the combat zone) which contains the lines of communications, establishments for supply and evacuation, and other agencies required for the immediate support and maintenance of the field forces.

NATO - The area of a theatre of operations immediately to the rear of the area of combat operations, that contains lines of communication, rear command and control organizations, logistic units and agencies, army group and theatre headquarters reserves.

Complex emergency:
UN, NATO, USAID - As defined by the UN Inter-Agency Standing Committee (IASC), this is "a humanitarian crisis in a country, region, or society where there is total or considerable breakdown of authority resulting from internal or external conflict, and which requires an international response that goes beyond the mandate or capacity of any single and/or ongoing UN country program."

USAID - 2. Natural or man-made disaster with economic, social, and political dimensions. A profound social crisis in which a large number of people die and suffer from war, disease, hunger, and displacement owing to man-made and natural disasters, while some others may benefit from it. Four factors can be measured: the fatalities from violence; the mortality of children under five years of age; the percentage of underweight children under five; and the number of external refugees and IDPs.

Conflict mitigation:
USAID - 1. Efforts to contain and reduce the amount of violence used by parties in violent conflict and engage them in a process to settle the dispute and terminate the violence.

USAID - 2. The reduction or minimization of violent acts normally targeted toward a specific group to compel restraint and restore calm.

Conflict prevention:
NATO - 1. The timely recognition and elimination of possible causes of conflict, implying the use of a range of political, diplomatic, military and other methods at early stages in its development.

NATO - 2. Activities ranging from diplomatic initiatives to preventive deployment of troops, intended to prevent disputes from escalating into armed conflicts or from spreading. They can include fact-finding missions, consultation, warnings, inspections, and monitoring.

Contingency:
DOD - An emergency involving military forces caused by natural disasters, terrorists, subversives, or by required military operations. Due to the uncertainty of the situation, contingencies require plans, rapid response, and special procedures to ensure the safety and readiness of personnel, installations, and equipment.

Contingency plan(ning):
UN - It involves preparing likely courses of action dealing with a range of potential scenarios and extends into preparatory activities (preparation of maps, identification of sources of equipment and supplies, pre-positioning of communications and identification of possible troop contributing states).

Crisis management:
DOD - Measure to resolve a hostile situation and investigate and prepare a criminal case for prosecution under federal law. Crisis management will include a response to an incident involving a weapon of mass destruction (WMD), special improvised explosive device (IED), or a hostage crisis that is beyond the capability of the lead federal agency.

NATO - The coordinated actions taken to defuse crises, prevent their escalation into an armed conflict and contain hostilities if they should result.

USAID - Efforts to keep situations of high tension and confrontation from breaking into armed violence, usually involving threats of force.

Crisis response operation (CRO):
NATO - Any NATO military operation undertaken for purposes of crisis management.

Demilitarization:
NATO - Demilitarization means that military personnel and equipment are withdrawn from their military function.

Demilitarized zone:
DOD, NATO - A defined area in which the stationing, or concentrating of military forces, or the retention or establishment of military installations of any description, is prohibited.

Demobilization:
DOD - The process of transitioning a conflict or wartime military establishment and defense based civilian economy to a peacetime configuration while maintaining national security and economic vitality.

NATO - Demobilization consists of those activities that are undertaken by a Peace Support Force to reduce the number of factions' forces and their equipment in the area of operations to the levels as agreed in the peace settlement.

Development:
USAID - 1. Long-term development efforts aimed at bringing improvements in economic, political, and social status and the quality of life of all segments of the population as well as environmental sustainability.

USAID - 2. Broad-based sustainable development has four components. The first is a healthy, growing economy that constantly transforms itself to maintain and enhance the standard of living. Second, the benefits of economic growth are equitably shared; women, minorities, immigrants, the poor, and the handicapped get a fair deal from economic growth. The third component includes respect for human rights, good governance, a vibrant civil society of NGOs, and an increasingly democratic society. The fourth is sustainability, which means that in the process of economic growth, we do not destroy the environment, enabling our descendants to enjoy the same or higher standard of living. (Weaver et. al 1997: 2-3)

Developmental assistance:
DOD - USAID function chartered under chapter one of the Foreign Assistance Act of 1961, primarily designed to promote economic growth and the equitable distribution of its benefits.

Direct assistance:
DOD - Face-to-face distribution of goods and services.

Disarmament:
NATO - A sub-process of demilitarization. It means the (controlled process) of taking weapons away from military forces. Demilitarization and disarmament usually take place within the framework of demobilization operations.

Disaster control:
DOD - Measures taken before, during, or after hostile action or natural or manmade disasters to reduce the probability of damage, minimize its effects, and initiate recovery.

Disaster relief:
NATO - Assistance and/or intervention during or after disaster to meet the life preservation and basic subsistence needs. It can be of emergency or protracted duration.

Disaster response:
NATO - A sum of decisions and actions taken during and after disaster, including immediate relief, rehabilitation, and reconstruction.

Emergency relief operation:
NATO - Emergency rescue and other urgent work in response to an emergency, aimed at preserving life and health, reducing environmental damage and material losses, and at containing the emergency.

End state:
DOD - The set of required conditions that defines achievement of the commander's objectives.

NATO - The political and/or military situation to be attained at the end of an operation, which indicates that the objective has been achieved.

Foreign emergency support team:
U.S. Marine Corps (USMC) - In consequence management (CM) scenarios involving intentional/malevolent use of WMD or CBRN material contamination, the DOS Office of Counter Terrorism deploys a Foreign Emergency Support Team (FEST) that gives the Ambassador robust communication and other capabilities and a 24-hour Command Center is staffed by the departments of Defense, Health and Human Services, Energy, Justice, and State, and by the FBI. The team has scientific assets that help preserve evidence.

Foreign humanitarian assistance:
USMC - Operations conducted to relieve or reduce the results of disaster brought on by either natural (flood, drought, fire, hurricane) or manmade (civil violence, nuclear, biological or chemical accident) causes, or other endemic conditions such as human pain, disease, hunger or privation in countries or regions outside the United States. It is generally limited in scope and duration; it is intended to supplement or complement efforts of the host nation civil authorities or agencies with primary responsibility for providing assistance.

Human development:
UN - The process of expanding people's choices.

Humanitarian action:
NATO - The work of IGOs and NGOs to ensure the survival or relief of civilians affected by military action in various types of conflict.

Humanitarian aid:
NATO - Measures taken to relieve civilian hardship, especially when local authorities are unable to provide the population with staple supplies or make no effort to do so. These measures may be part of a peace support operation or an independent program.

Humanitarian assistance:
NATO - Missions conducted to relieve human suffering, especially in circumstances where responsible authorities in the area are unable, or possibly unwilling, to provide adequate service support to the population. Humanitarian aid missions may be conducted in the context of a peace support operation, or as a completely independent task.

UK DFID Assistance comprises disaster relief, food aid, refugee relief, and disaster preparedness. It generally involves the provision of material aid including food, medical care, and personnel and finance and advice to save and preserve lives during emergency situations and in the immediate post- emergency rehabilitation phase; and to cope with short and longer term population displacements arising out of emergencies

USMC - Programs conducted to relieve or reduce the results of natural or manmade disasters or other endemic conditions such as human pain, disease, hunger, or privation that might present a serious a serious threat to life or that can result in great damage to or loss of property. Humanitarian assistance provided by U.S. forces is limited in scope and duration. The assistance provided is designed to supplement or complement the efforts of the host nation civil authorities or agencies that may have the primary responsibility for providing humanitarian assistance.

Humanitarian assistance contingency:
DOD - A contingency resulting from natural or manmade disasters or other endemic conditions such as human pain, disease, hunger, or privation that might present a serious threat to life or that can result in great damage to or loss of property.

Humanitarian assistance coordination center (HACC):
DOD - A temporary center established by a geographic combatant commander to assist with interagency coordination and planning. A HACC operates during the early planning and coordination stages of foreign humanitarian assistance operations by providing the link between the geographic combatant commander and other USG agencies, NGOs, and international and regional organizations at the strategic level.

Humanitarian and civic assistance (HCA):
DOD - Assistance to the local populace provided by predominantly U.S. forces in conjunction with military operations and exercises. This assistance is specifically authorized by title 10, United States Code, section 401, and funded under separate authorities. Assistance provided under these provisions is limited to medical, dental, and

veterinary care provided in rural areas of a country; construction of rudimentary surface transportation systems; well drilling and construction of basic sanitation facilities; and rudimentary construction and repair of public facilities. Assistance must fulfill unit training requirements that incidentally create humanitarian benefit to the local populace.

Humanitarian assistance operation:
DOD - Military operations providing a secure environment to allow humanitarian relief efforts to progress and includes: Disaster Relief, Refugee Assistance, and Humanitarian and Civic Assistance and Civil Support.

Humanitarian interventions:
USAID - Reliance on force for the justifiable purpose of protecting the inhabitants of another state from treatment that is arbitrary and persistently abusive.

Humanitarian information center:
DOD - An interagency policymaking body that coordinates the overall relief strategy and unity of effort among all participants in a large foreign humanitarian assistance operation. It normally is established under the direction of the government of the affected country or the UN, or a USG agency during a U.S. unilateral operation. The humanitarian operations center should consist of representatives from the affected country, the U.S. Embassy or Consulate, the joint force, the UN, NGOs, IOs, and other major players in the operation.

Humanitarian emergency:
USAID - 1. Situations in which large numbers of people are dependent on humanitarian assistance from sources external to their own society and/or are in need of physical protection to have access to subsistence or external assistance.

USAID - 2. A profound social crisis in which a large number of people die and suffer from war, disease, hunger. and displacement owing to man-made and natural disasters, while some others may benefit from it.

Humanitarian operations:
NATO - An operation; usually conducted under the aegis of an IO, to prevent a humanitarian disaster and to provide humanitarian aid to the civilian population in an area of armed conflict.

Humanitarian operations center (HOC):
DOD - An interagency policymaking body that coordinates the overall relief strategy and unity of effort among all participants in a large foreign humanitarian assistance operation. It normally is established under the direction of the government of the affected country or the UN, or a USG agency during a U.S. unilateral operation. The humanitarian operations center should consist of representatives from the affected country, the U.S. Embassy or Consulate, the joint force, the UN, NGOs, IOs, and other major players in the operation.

Indirect assistance:
DOD - Assistance that is at least one step removed from direct contact with the affected population, often involving such activities as transporting relief goods or relief personnel to an affected area of operation.

Information management :
DOD - Efficient mechanisms for the collection, analysis, storage, and accessibility of information to achieve the optimal use of that information. This is a key component of civil-military coordination.

Information support:
NATO - The activities of specialized information-processing and analysis centers providing information for officials, agencies, and bodies responsible for planning and carrying out a peacekeeping operation and keeping the public informed on progress and results.

Infrastructure support:
DOD - Provision of general services, such as road repair, airspace management and power generation that facilitate relief, but are not necessarily visible to, or solely for, the benefit of the affected population.

Interagency community:
DOD - The broad interagency community can include U.S. Government departments; federal, state, or local agencies; foreign governments singularly or as an alliance or coalition; global and regional international or U.S. and foreign NGOs.

Interagency operations:
DOD - Any military or civilian activity, across the full spectrum of operations, in which government or non-governmental agencies communicate, coordinate, cooperate, collaborate, and/or integrate to achieve unity of effort.

Internally displaced person (IDP):
DOD - Any person who has left their residence by reason of real or imagined danger but has not left the territory of their own country.

NATO - A person who, as part of a mass movement, has been forced to flee his or her home or place of habitual residence suddenly or unexpectedly as a result of armed conflict, internal strife, systematic violation of human rights, fear of such violation, or natural or man-made disasters, and who has not crossed an internationally recognized state border.

UN - A person forced or obliged to flee or to leave their homes or places of habitual residence, in particular or as a result of or in order to avoid the effects of . . . violations of human rights. . . and who have not crossed an internationally recognized State border.

USAID - Individuals who have been forced to flee their homes for the same reasons as refugees but have not crossed an internationally recognized border.

Internal armed conflict:
NATO - A conflict within a state in which armed force is used. It may have its origin in the activities of extremist, nationalist, religious, separatist, or terrorist movements and groupings. Also called intra-state armed conflict.

Inter-governmental organizations (IGOs):
DOD - Bodies formed by national governments to establish cooperation on issues relating to economics, security, culture, politics, or common geographic concerns. IGOs are based on charters and treaties, governed by representatives of member governments, and receive funding support from their member nations. Examples include the UN, NATO, and the OSCE.

International organization (IO):
DOD - An organization with an international or global mandate, generally funded by contributions from national governments. Examples include the ICRC, the International Organization for Migration, and UN agencies.

Knowledge management:
DOD - The systematic process and strategy for finding, capturing, organizing, distilling, and presenting data, information, and knowledge for a specific purpose and to serve a specific organization or community. Knowledge management addresses the problem of accumulating an overload of data and information without systematic organization and synthesizing to allow presentation and sharing among collaborative partners.

Large-scale wars:
NATO - A war involving a substantial number of states from various regions in the world aimed at achieving fundamental politico-military objectives.

Limited military action:
NATO - The use by a state of military force limited in aims, area, and time against other countries or illegal armed formations. It is resorted to by the political and military leaders of a country when the ends pursued do not call for large-scale use of the armed forces, or when this is impossible or undesirable.

Limited war:
DOD - Armed conflict short of general war, exclusive of incidents, involving the overt engagement of the military forces of two or more nations.

Low intensity conflict:
DOD, USAID - Political-military confrontation between contending states or groups below conventional war and above the routine, peaceful competition among states. It frequently involves protracted struggles of competing principles and ideologies. Low intensity conflict ranges from subversion to the use of armed force. It is waged by a

combination of means employing political, economic, informational, and military instruments. Low intensity conflicts are often localized, generally in the Third World, but contain regional and global security implications.

USAID - 2. Conflict involving armed combat or acts of terrorism on a protracted but sporadic basis.

Military operations other than war (MOOTW):
DOD, UN - Operations that encompass the use of military capabilities across the range of military operations short of war. These military actions can be applied to complement any combination of the other instruments of national power and occur before, during, and after war.

USAID - The range of military actions required by the President, except those associated with major combat operations conducted pursuant to a declaration of war or authorized by the War Powers Limitation Act or a joint resolution of Congress in support of national security interests and objectives. These military actions can be applied to complement any combination of the other instruments of national power.

Minor armed conflict:
USAID - Violent conflict in which the total number of battle-related deaths during the course of the conflict is below 1,000.

Non-governmental organizations (NGOs):
DOD - Private-sector organizations established for many purposes and often specializing in narrowly defined tasks for specific religious, social, charitable, or political purposes. They generally are formed under the laws of a specific nation, are directed by private citizens, and do not generate profits from their activities. Examples include CARE, Oxfam, and World Vision.

Peace operations (PO):
DOD - A broad term that encompasses peacekeeping operations and peace enforcement operations conducted in support of diplomatic efforts to establish and maintain peace.

UN - peace support operations, includes: preventive deployments; peacekeeping and peace-enforcement operations; diplomatic activities, such as preventive diplomacy, peacemaking, and peace building; as well as humanitarian assistance, good offices, fact-finding, and electoral assistance.

USAID - The umbrella term which encompasses three types of activities: activities with predominately diplomatic orientation, (preventive diplomacy, peacemaking, peace building); and two complementary, predominately military activities, namely peacekeeping and peace enforcement.

Peace support operations:
NATO - Any NATO military operation undertaken in support of peacekeeping. A peace support operation falls under the category of crisis response operations, just as peacekeeping is one aspect of crisis management.

Peacekeeping:
DOD - Military operations undertaken with the consent of all major parties to a dispute, designed to monitor and facilitate implementation of an agreement (ceasefire, truce, or other such agreement) and support diplomatic efforts to reach a long-term political settlement.

NATO - The containment, moderation and/or termination of hostilities between or within states, through the medium of an impartial third party intervention, organized and directed internationally; using military forces, and civilians to complement the political process of conflict resolution and to restore and maintain peace.

UN - PK Hybrid politico-military activity aimed at conflict control, which involves a UN presence in the field (usually military and civilian personnel), with the consent of the parties, to implement or monitor the implementation of arrangements relating to the control of conflicts (cease-fires and separation of forces), and their resolution (partial or comprehensive settlements) and/or to protect the delivery of humanitarian relief.

USAID - Neutral military or paramilitary operations undertaken with the consent of all major belligerents, designed to monitor and facilitate implementation of existing truces and support diplomatic efforts to reach a long-term political settlement.

Peacekeeping operations:
UN - PKO Non-combat military operations undertaken by outside forces with the consent of all major belligerent parties and designed to monitor and facilitate the implementation of an existing truce agreement in support of diplomatic efforts to reach a political settlement; 'PKOs' covers: peace-keeping forces, observer missions and mixed operations.

USAID - A common term used for various types of activities, such as to resolve conflict; prevent conflict escalation; halt or prevent military actions; to uphold law and order in a conflict zone; conduct humanitarian actions; restore social and political institutions whose functioning has been disrupted by the conflict; and restore basic conditions for daily living. The distinctive feature of peacekeeping operations is that they are conducted under a mandate from the UN or regional organizations whose functions include peace support and international security.

Peace-making:
DOD - The process of diplomacy, mediation, negotiation, or other forms of peaceful settlements that arranges an end to a dispute and resolves issues that led to it.

NATO - Diplomatic actions conducted after the commencement of conflict, with the aim of establishing a peaceful settlement. They can include the provision of good offices, mediation, conciliation, and such actions as diplomatic isolation and sanctions.

UN - Diplomatic process of brokering an end to conflict, principally through mediation and negotiation, as foreseen under Chapter VI of the UN Charter; military activities contributing to peacemaking include military-to-military contacts, security assistance, shows of force and preventive deployments.

USAID - A process of diplomacy, mediation, negotiation, or other forms of achieving peaceful settlements that arrange ends to disputes.

Peace building:
DOD - Post-conflict actions, predominately diplomatic and economic, that strengthen and rebuild governmental infrastructure and institutions in order to avoid a relapse into conflict.

NATO - Post-conflict action to identify and support structures which will tend to strengthen and consolidate a political settlement in order to avoid a return to conflict. It includes mechanisms to identify and support structures which will tend to consolidate peace, advance a sense of confidence and wellbeing and support economic reconstruction, and may require military as well as civilian involvement.

UN - In the aftermath of conflict; it means identifying and supporting measures and structures which will solidify peace and build trust and interaction among former enemies, in order to avoid a relapse into conflict; often involves elections organized, supervised or conducted by the UN, the rebuilding of civil physical infrastructures and institutions such as schools and hospitals, and economic reconstruction.

USAID - 1. The employment of measures to consolidate peaceful relations and create an environment that deters the emergence or escalation of tensions which may lead to conflict.

USAID - 2. The effort to promote human security in societies marked by conflict. The overarching goal of peace building is to strengthen the capacity of societies to manage conflict without violence, as a means to achieve sustainable human security.

Peace enforcement:
DOD - Application of military force, or the threat of its use, normally pursuant to international authorization, to compel compliance with resolutions or sanctions designed to maintain or restore peace and order.

USAID - 1. The application of military force or threat of its use, normally pursuant to international authorization, to compel compliance with generally accepted resolutions or sanctions to maintain or restore peace and support diplomatic efforts to reach a long-term political settlement. The primary purpose of peace enforcement is the restoration of peace under conditions broadly defined by the international community.

USAID - 2. The use or threat of armed force as provided for in Chapter VII of the UN Charter aimed at restoring peace by military means such as in Korea (1950-1953) or Iraq (1991). It can take place without the agreement and support of one or all of the warring parties. It can refer to both an inter-state or an intrastate conflict, to [serve] the mitigation of a humanitarian emergency or in situations where the organs of state have ceased to function. Peace enforcement actions include carrying out international sanctions against the opposing sides, or against the side that represents the driving force in the armed conflict; isolating the conflict and preventing arms deliveries to the area, as well as preventing its penetration by armed formations; delivering air or missile strikes on positions of the side that refuses to halt its military actions; and rapid deployment of peace forces to the combat zones in numbers sufficient to carry out the assigned missions, including the localizing of the conflict and the disarming or eradicating of any armed formations that refuse to cease fighting.

Reconstruction:
USAID - 1. The permanent reconstruction or replacement of severely damaged physical structures, the full restoration of all services and local infrastructure, and the revitalization of the economy.

USAID - 2. Economic, political and social rebuilding of post-conflict state and society, including de-mining, disarmament, reintegration of combatants, return of refugees, resettlement of internally displaced persons, reviving political processes, restoring physical infrastructures, re-starting economic life, conversion to civilian production, re-establishing civilian authority, and conducting new (supervised) elections.

Refugee:
DOD - A person who, by reason of real or imagined danger, has left their home country or country of their nationality and is unwilling or unable to return.

NATO - Any person who, owing to a well-founded fear of being persecuted for reasons of race, religion, nationality, membership of a particular social group or political opinion, is outside the country of his nationality and is unable, or owing to such fear is unwilling, to avail himself of the protection of that country; or who, not having a nationality and being outside the country of his former habitual residence as a result of such events, is unable, or owing to such fear is unwilling, to return to it.

USAID - 1. A person who, owing to a well-founded fear of being persecuted for reasons of race, religion, nationality, or membership of a particular social group or political opinion, is outside the country of his nationality and is unable or, owing to such fear, is unwilling to avail himself of the protection of that country.

USAID - 2. Every person who, owing to external aggression, occupation, foreign domination or events seriously disturbing public order in either part or the whole of his country of origin or nationality, is compelled to leave his place of habitual residence in order to seek refuge in another place outside his country of origin or nationality.

Sector:
DOD, NATO - An area designated by boundaries within which a unit operates, and for which it is responsible.

UN - A functional area, such as water/sanitation, food, or shelter.

Small war:
NATO - A range of ancillary, often improvised actions designed to inflict losses on an enemy wherever possible and by all available means in order to obtain decisive results on the main fronts in an armed struggle.

Stability operations:
DOD - Military and civilian activities conducted both in peacetime and across the full spectrum of conflict to establish order in failed or failing states and regions.

Sustainable development:
USAID - Continued economic and social progress that rests on four key principles: improved quality of life for both current and future generations; responsible stewardship of the natural resource base; broad-based participation in political and economic life; and effective institutions which are transparent, accountable, responsive, and capable of managing change without relying on continued external support. The ultimate measure of success of sustainable development programs is to reach a point where improvements in the quality of life and environment are such that external assistance is no longer necessary and can be replaced with new forms of diplomacy, cooperation, and commerce.

U.S. government interagency process:
DOD - A national strategic mechanism designed to ensure information and options are developed and passed up to the national leadership, so decisions and guidance can be passed down to the appropriate level staffs, which can write the orders and oversee their execution.

Wider peacekeeping:
UK - The wider aspects of peacekeeping operations carried out with the consent of the belligerent parties but in an environment that may be highly volatile.

Appendix B: Internet Web Sites

Adventist Development and Relief Agency International http://www.adra.org

Afghanistan Information Management Service (AIMS) http://www.aims.org.af

African Center for the Constructive Resolution of Disputes http://www.accord.org.za

AidMatrix http://www.aidmatrix.org

Aid Workers Network http://www.aidworkers.net/

Aid World http://www.aidworld.org

Alertnet http://www.alertnet.org

American Council for Voluntary International Action (InterAction) http://www.interaction.org/

American Red Cross http://www.redcross.org/

American Jewish World Service http://www.ajws.org/

Amnesty International http://www.amnesty.org/

Army Corps of Engineers http://www.usace.army.mil

Baptist World Alliance http://www.bwanet.org

CACI International Inc. http://www.caci.com/

Camber Corporation http://www.camber.com/about.asp?l=b

Canadian Center for Emergency Preparedness http://www.ccep.ca

CARE http://www.care.org/

Caribbean Disaster Emergency Response Agency http://www.cdera.org

Catholic Relief Services http://www.catholicrelief.org/

Center for Humanitarian Cooperation http://www.cooperationcenter.org

Center for International Development and Conflict Management http://www.cidcm.umd.edu

Center for Research on the Epidemiology of Disasters http://www.cred.be

Center for Stabilization and Reconstruction Studies, Naval Post-Graduate School
http://www.nps.edu/csrs

Center for Strategic and International Studies http://www.csis.org/

National Defense University, Center for Technology & National Security Policy
http://www.ndu.edu/ctnsp/S&R_IT_workshop.htm

Center of Excellence in Disaster Management and Humanitarian Assistance
http://coe-dmha.org

Centers for Disease Control and Prevention http://www.cdc.gov/

Church World Service http://www.churchworldservice.org/index.html

CISCO Systems, Inc. http://www.cisco.com/

Civil Military and Coordination Section http://ochaonline.un.org/mcdu

Civil Protection within the European Countries http://europa.eu.int/comm/environment/civil/

Civilian and Military Cooperation in Complex Humanitarian Operations, by Sarah E. Archer
http://www.army.mil/professionalwriting/volumes/volume1/june_2003/6_03_2.html

Collaborative Learning Projects and the Collaborative for Development Action, Inc.
http://www.cdainc.com

Cranfield Disaster Management Center http://www.rmcs.cranfield.ac.uk/dmc/ddmsa/dmc

Crisis Management Initiative http://www.cmi.fi/?content=itcm_project

Current Peacekeeping/Building Missions
http://www.un.org/Depts/dpko/dpko/cu_mission/body.htm

Defense Information Systems Agency http://www.disa.mil

DFI International http://www.dfi-intl.com/

Direct Relief International http://www.directrelief.org/

Disaster Management Center, University of Wisconsin/Madison http://dmc.engr.wisc.edu/about/

Disaster Management Institute of South Africa http://www.disaster.co.za/index.php

DisasterRelief http://www.disasterrelief.org/

Disaster Research Center http://www.udel.edu/DRC/

Disaster Relief And Strategic Telecommunications Infrastructure Company
http://www.drasticom.org

E-Centre http://www.the-ecentre.net/

Emergency Preparedness Information Exchange http://epix.hazard.net

Ericsson Response Program http://www.ericsson.com

European Community Humanitarian Office http://europa.eu.int/comm/echo/en/index_en.htm

Federal Emergency Management Agency http://www.fema.gov

International Alert http://www.international-alert.org/

Food and Agriculture Organization of the United States http://www.fao.org

Foundation Hirondelle http://www.hirondelle.org

Fritz Institute http://www.fritzinstitute.org

FSI International http://www.fsi-intl.com/

Geographic Information Support Team http://www.gist.itos.uga.edu

Geneva Center for Security Policy (GCSP) http://www.gcsp.ch/e/index.htm

Geneva Center for the Democratic Control of Armed Forces (DCAF) http://www.dcaf.ch

Global Hand http://www.globalhand.org

Global Impact http://www.charity.org/

Global Internally Displaced Persons Project http://www.idpproject.org

Global Map Aid http://www.globalmapaid.rdvp.org

GlobalsSecurity.org http://www.globalsecurity.org/

Global Relief Technologies http://www.globalrelieftech.com/

Global Strategies Group http://www.gsghq.com/

Grassroots International http://www.grassrootsonline.org/

Groove Networks virtual office http://www.groove.net/home/index.cfm

Gurtong Peace Project http://www.gurtong.org

Habitat for Humanity International http://www.habitat.org/

Headquarters, Department of the Army http://www.hqda.army.mil/hqda/main/home.asp

HumaniNet http://www.humaninet.org

Humanitarian Information Centers and Partners http://www.humanitarianinfo.org

Humanitarian Information Unit http://hiu.state.gov/

Humanitarian Early Warning Service http://www.hewsweb.org

Humanitarian Policy and Conflict Research http://www.hpcr.org

Human Rights Watch http://www.hrw.org/

I-LINX http://www.i-linx.net

Information Management for Humanitarian Operations http://www.currion.net/imho.htm

Information Technology and Crisis Management http://www.itcm.org/

InfoShare http://www.info-share.org

Institute for Defense Analysis http://www.ida.org/

Integrated Regional Information Network http://www.irinnews.org

Interagency Transformation, Education & After Action Review http://www.ndu.edu/itea

International Aid http://www.internationalaid.org

International Institute for Disaster Risk Management http://www.idrmhome.org

International Committee of the Red Cross and Red Crescent http://www.icrc.org

International Council of Voluntary Agencies http://www.icva.ch/

International Crisis Group http://www.crisisgroup.org/home/index.cfm

International Humanitarian Law and Research Initiative http://www.ihlresearch.org/iraq/

International Institute for Democracy and Electoral Assistance http://www.idea.int/index.cfm

International Institute for Disaster Risk Management http://www.idrmhome.org

International Justice Mission http://www.ijm.org

International Organization for Migration http://www.iom.int

International Orthodox Christian Charities http://www.iocc.org

International Search and Rescue Advisory Group http://www.reliefweb.int/insarag/

International Rescue Committee http://www.theIRC.org

International Relief and Development, inc. http://www.ird-dc.org/

International Telecommunication Union http://www.itu.int

Inter-Agency Standing Committee http://www.humanitarianinfo.org/iasc/

Inveneo http://www.inveneo.org/

Islamic Relief http://www.islamic-relief.com/

Journal of Humanitarian Assistance http://www.jha.ac/

Luise Druke http://www.luisedruke.com

Lutheran Relief Services http://www.lwr.org/

Map Action http://www.mapaction.org

MapRelief http://www.maprelief.org

Martus Technology Non-Profit http://www.martus .org

Médecins Sans Frontières http://www.doctorswithoutborders.org/

Mennonite Central Committee http://www.mcc.org/

Mercy Corps International http://www.mercycorps.org

MorganFranklin Corporation http://www.morgan-franklin.com/

National Defense University, Institute for National Strategic Studies http://www.ndu.edu/inss/

National Democratic Institute for International Affairs http://www.ndi.org

National Geospatial Intelligence Agency http://www.nga.mil

National Strategic Gaming Center http://www.ndu.edu/inss/nsgc

NATO/Military Acronyms http://www.nato.int/ifor/general/acronyms.htm

Naval Air Systems Command http://www.navair.navy.mil/

Navy Construction Forces http://www.seabee.navy.mil

Naval Postgraduate School, Center for Stabilization and Reconstruction Studies
http://www.nps.edu/CSRS/

Naval War College http://www.nwc.navy.mil/defaultf.htm

NetHope http://www.nethope.org/

Network Startup Research Center http://www.nsrc.org/

Nonprofit Technology Enterprise Network http://www.nten.org/

Object Management Group http://www.omg.org/

Office of the Assistant Secretary of Defense for Special Operations and Low-Intensity Conflict
http://www.defenselink.mil/policy/solic

Office of the Assistant Secretary of Defense for Networks and Information Integration
http://www.defenselink.mil/nii

Office for the Coordination of Humanitarian Affairs http://www.reliefweb.int/symposium/

OneWorld.net http://www.OneWorld.net

Organization Economic Co-operation and Development http://www.oecd.org/

Overseas Development Institute http://www.odi.org.uk/about.html

OXFAM International http://www.oxfam.org/

Pacific Disaster Center http://www.pdc.org

Pan American Health Organization http://www.paho.org

Partnership for Peace Information Management System http://www.pims.org

Partners in Technology http://www.pactec.org

Peacekeeping and Stability Operations Institute (PKSOI), US Army War College
http://www.carlisle.army.mil/usacsl/divisions/pksoi/default.htm

Prevention Consortium http://www.proventionconsortium.org

Proactive Communications Inc. http://204.200.201.151/

Refugees International http://www.refugeesinternational.org/

Relief Guide http://www.reliefguide.org

Reliefweb http://www.reliefweb.int

RESPOND http://www.respond-int.org

Responsibility to Protect, "Report of the International Commission on Intervention and State Sovereignty" http://www.iciss.ca/report-en.asp

SAIC http://www.saic.com/

Sandia National Laboratories http://www.sandia.gov/

Save the Children http://www.savethechildren.org/

Search for Common Ground http://www.sfcg.org/

Standardized Monitoring and Assessment of Relief and Transition http://www.smartindicators.org

Status of Forces Agreement http://www.globalsecurity.org/military/facility/sofa.htm

Stockholm International Peace Research Institute (SIPRI) http://www.sipri.org

Swiss Peace http://www.swisspeace.org

Swiss Seismological Service http://www.seismo.ethz.ch

Tele Communication Systems http://www1.telecomsys.com/index.html

Télécoms Sans Frontières http://www.tsfi.org/

Thought Link http://www.thoughtlink.com/

Training Transformation http://www.t2net.org/

UK Post Conflict Reconstruction Unit http://www.postconflict.gov.uk/

UK Department for International Development http://www.dfid.gov.uk/

UN System http://www.unsystem.org

United Nations Non-Governmental Liaison Service (NGLS) http://www.un-ngls.org/

UN Disaster Management Training Program http://www.undmtp.org

UN Disaster Assessment and Coordination http://www.reliefweb.int/undac/

UNICEF Humanitarian Principles Training, On-line session http://coe-dmha.org/Unicef/UNICEF2FS.htm

United Nations Children Fund http://www.unicef.org

United Nations Development Program http://www.undp.org.in/

United Nations High Commissioner for Refugees http://www.unhcr.ch

United Nations Home Page http://www.un.org

United Nations Joint Logistics Center http://www.unjlc.org

UN Office for the Coordination of Humanitarian Affairs http://ochaonline.un.org

UN Satellite Imagery Organization http://www.unosat.org

US Agency for International Development http://www.usaid.gov

US Air Force www.airforce.com/

US Army Peacekeeping & Stability Operations Institute
http://www.carlisle.army.mil/usacsl/divisions/pksoi/Welcome/welcomebody.htm

US Army Civil Affairs and PsyOps Command http://www.usacapoc.army.mil

US Army War College http://www.carlisle.army.mil/

US Institute of Peace www.usip.org

US Department of State http://www.state.gov

US Department of State Office of the Coordinator for Reconstruction and Stabilization
http://www.state.gov/s/crs/

US Geological Survey http://www.usgs.gov

US Marine Corps www.usmc.mil/

US Navy http://www.navy.mil/

US Office of Foreign Disaster Assistance http://www.usaid.gov/hum_response/ofda

US Joint Forces Command http://www.jfcom.mil

US Northern Command http://www.northcom.mil

US Southern Command http://www.southcom.mil

US European Command http://www.eucom.mil

US Central Command http://www.centcom.mil

US Pacific Command http://www.pacom.mil

US Special Operations Command http://www.socom.mil

Virtual Operations On-Site Coordination Center http://ocha.unog.ch/virtualosocc/

War-Torn Societies Project http://www.wsp-international.org

World Food Program http://www.wfp.org

World Health Organization http://www.who.int

World Relief http://www.wr.org/

World Vision http://www.worldvision.org

Appendix C: Acronyms

AOR		Areas of Responsibility
ASEAN		Association of Southeast Asian Nations
AU	African	Union
AUS AID		Australian Agency for International Development
CARE		Cooperative Assistance for Relief Everywhere
CCP	Classified	Connectivity Program
CENTRIXS		Coalition Enterprise Regional Information Exchange System
CEPREDENAC		Coordination Center for the Prevention of Natural Disaster in Central America
CIDA		Canadian International Development Agency
CIE		Collaborative Information Environment
CIMIC		Civil-Military Coordination
CIO		Chief Information Officer
CIP	Critical	Infrastructure Protection
CJTF		Combined Joint Task Force
CMIC		Civil-Military Cooperation Center
COI		Community of Interest
CONOPS		Concept of Operations
CIR		Committed Information Rate
CITEL		Inter-American Telecommunications Commission
CMO	Civil-M	ilitary Operation
CMOC		Civil-Military Operations Center
COW		Cellular on Wheels
CRICOM		Caribbean Community and Common Market
CRO		Crisis Response Operation
CRS	Catholic	Relief Service
CSD		United Nations Commission on Sustainable Development
CSMP		Contingency Support and Migration Planning
CTNSP		Center for Technology and National Security Policy
CWS	Church	World Services
DA	Developm	ent Assistance
DART		Disaster Assistance Response Team
DFID		UK Department for International Development
DIA		Defense Intelligence Agency
DOD		U.S. Department of Defense
DRC	Danish	Relief Council
EADRCC		Euro-Atlantic Disaster Response Coordination Center
EADRU		Euro-Atlantic Disaster Response Unit
ECHO		European Community Humanitarian Aid Office
ECOWAS		Economic Community Of West African States
EIPC	Enhanced	International Peacekeeping Capabilities
ERMA		Emergency Refugee & Migration Assistance
ESDI	European	Security and Defense Initiative
ESDP		European Security and Defense Program

EU	European	Union
FCC	Federal	Communications Commission
FEMA		Federal Emergency Management Agency
FEST		Foreign Emergency Support Team
FISMA		Federal Information Security Management Act
FPC	Foreign	Press Center
GDIN		Global Disaster Information Network
GDP		Gross Domestic Product
GIS		Geographic Information System
GNP		Gross National Product
GPS		Global Positioning System
GRT		Global Relief Technologies
HADR		Humanitarian Assistance and Disaster Relief
HIC		Humanitarian Information Center
HMA		Humanitarian Mine Action
IC	Inte	lligence Community
ICRC		International Committee of the Red Cross
ICT		Information and Communications Technology
ICVA		International Council of Voluntary Agencies
IFRC		International Federation of Red Cross and Red Crescent Societies
ILMS		Integrated Logistics Management System
IGO		Inter-Governmental Organization
IMF		International Monetary Fund
INS		Immigration and Naturalization Service
IO		International Organization
IOC		Initial Operating Capability
IOM	International	Organization for Migration
IMC		International Medical Corps
IMF		International Monetary Fund
IP	Intern	et Protocol
IRC	Intern	ational Rescue Committee
IT	Inform	ation Technology
ITU	International	Telecommunication Union
KM	Knowledge	Management
LIC		Low Intensity Conflict
LISH		Logistics Information Security and Health
LOC		Lines of Control
LROBP		Long-Range Overseas Buildings Plan
MCI		Mercy Corps International
MOOTW		Military Operations Other Than War
MSF		Doctors without Borders
NATO		North Atlantic Treaty Organization
NATMC		NATO Air Traffic Management Committee
NGO		Non-Governmental Organization
NCA		National Command Authorities
NDU		National Defense University

NGA		National Geospatial Intelligence Agency
NIACAP		National Information Assurance Certification and Accreditation Program
NISTCAP		National Institute of Standards and Technology Certification and Accreditation Program
NIST		National Institute of Standards and Technology
NRC		Norwegian Relief Council
NSA	National	Security Agency
NSC		National Security Council
NSF	National	Science Foundation
NSTISSI		National Security Telecommunications and Information Systems Security Instruction
NTIA	National	Telecommunications and Information Administration
OAS		Organization of American States
OASD NII		Office of the Assistant Secretary of Defense Networks and Information Integration
OOTW		Operations Other Than War
OSAC		Overseas Security Advisory Councils
OSCE		Organization for Security and Cooperation in Europe
OSIS		Open Sources Information System
OSTP		Office of Science and Technology Policy
OXFAM		Oxford Committee for Famine Relief
PCRU		Post-Conflict Reconstruction Unit
PDA	Personal	Digital Assistant
PHS		Public Health Services
PKI		Public Key Infrastructure
PKO		Peacekeeping Operations
PMC		Private Military Company
PrepCom		Preparation Communications
PSMC		Peace Support Missions Concluded
PSTN		Public Switched Telephone Network
QoS		Quality of Service
ROE		Rules of Engagement
S&R		Stability and Reconstruction
S&T	Science	and Technology
SCI		Secure Compartmentalized Information
SENTRI		Secure Electronic Network for Travelers' Rapid Inspection
SHAPE		Supreme Headquarters Allied Powers Europe
SIPRNET		Secret Internet Protocol Router Network
SMART		State Messaging and Archive Retrieval Toolset
SMOM		Sovereign Military Order of Malta
SNAP	Spouse	Networking Assistance Program
SOP		Standard Operating Procedures
SSPP		Systems Security Program Plan
TC IAEA		Technical Cooperation Programs
TSF		Télécoms Sans Frontières

TSU	Technical	Security Upgrade
TSWG		Technical Support Working Group
TWG	Technical	Working Group
UMCOR		United Methodist Committee on Relief
UNDP		UN Development Program
UNDPKO		United Nations Department of Peacekeeping Operations
UNHCR		United Nations High Commissioner for Refugees
UNHCR		United Nations High Commissioner for Refugees
UNICEF		United Nations Children's Fund
UNSCR		United Nations Security Council Resolution
USAID		United States Agency for International Development
USCG		United States Coast Guard
USINFO		United States Information
UTM		Universal Transverse Mercator
VITA		Volunteers in Technical Assistance
VPN	Virtua	l Private Network
VSAT		Very Small Aperture Terminal
WB		World Bank
WFP		World Food Program
WHO		World Health Organization

Appendix D

Civil-Military Interaction Advice from Strong Angel
Eric Rasmussen, MD, FACP
Fleet Surgeon, Third Fleet
27 June 2000 Rim of the Pacific 2000

The summary thoughts below have been very liberally borrowed from smart people. Some are previously published, some are original statements from Strong Angel participants, and some are just our discussed and considered opinion of a productive way to do things. Few are likely to be innovative. It has been found though interviews that lessons learned the hard way tend to repeat themselves with no prompting.

For reference purposes, there are three general categories of civil-military interaction:
- Conflict management (both intervention and post-conflict rebuilding)
- Natural disaster response
- Complex emergency support

Ten Commandments

1) The military should generally not be in overall charge. The JTF Commander should be clearly subordinate to civilian authorities whenever possible. This:
- Sets a democratic process in place that's reassuring to the population served
- Sets expectations for the levels of responsibility
- Keeps the military footprint to an absolute minimum
  - With a transition requirement apparent daily
- Allows some mission drift (a desirable flexibility) while minimizing mission creep
- Maintains a coordinated Host Nation Support mandate
  - Establish the scope of that mandate early and often

2) Technology cannot substitute for personal interaction:
- Use all available modes of communication
  - Decide in common when and how to use technology
  - Don't assume face-to-face is ideal, but it's a good default
- Agree early on common definitions of important terms
- Do not assume that each understands the other. Cultural differences can be subtle, but profound
- Get out and talk with counterparts frequently. Share food and drink, equipment and resources. It's proven to save lives.

3) Personalities are more important than processes:
- Don't throw away the book; thoughtful and experienced people wrote it, but be able to flex. You're the one on the scene. Value your own leadership in context.

136

4) Know the cultures and issues that surround you:
- Avoid imposing your standards and beliefs, but remember that there are international declarations defining fundamental human rights and we've subscribed to them.
- Avoidable misunderstandings can cause distracting escalations.
- Strive for impartiality to the sides within the conflict.
- Breadth of understanding fosters a better recognition of the scope of your real task. Educate your people.

5) Work on building communications networks as you begin to plan:
- Key people need to communicate early across organizations. Make sure the Civ can talk to the Mil, the Mil to the Civ, and that both need to frequently.
- Provide power where it's needed.
- Get a multi-pathway phone book out and keep it updated.
- If you can't communicate, you can't coordinate.

6) Centralize planning and de-centralize execution:
- Trust your people, but be thoughtful about whom you empower.

7) Coordinate everything with everybody to the greatest possible extent:
- Few issues will doom a civil-military mission more quickly than the perception of arrogance on either side. Both sides employ professionals. Both sides have agendas. Good work is still possible.
- The troops need to know their role in the larger picture. Again, educate your people.

8) Remember that the UN agencies are distinct, anarchic, and highly effective:
- UN agencies agree through collaboration and consensus. Don't try to impose force.
- UN agency technical communications structure has often been the best in the area.
- The UN is no monolith. It is more than 50 agencies with various management structures, but a common theme is that the management is often very flat; occasionally directly from highest headquarters to a local representative in the field.
- UN agencies consider their mandate to be externally focused while we often look at "self" and "other," with an internal focus. Not necessarily wrong either way, but very different. This is worth remembering.
- UN agencies were there before, are there now, and will be there afterwards. They know a lot about the neighborhood and they understand the situation. They know we're going to leave as soon as we can and that affects our interaction.
- UN agencies look to the military to provide four areas of support: "LISH"
  - Logistics support (particularly heavy lift)
  - Information (particularly local conditions)

- Security (of UN and other non-governmental staff)
- Health (primary care for the UN staff)
    - The key to success will be coordinating and cooperating

9) Senior Commanders and Staffs need education and training for non-traditional roles:
- Troops need awareness and understanding. Push the situational awareness out as far as you can. Educate your people.

10) Even in a seemingly simple operation there WILL BE more media and more politics than anticipated. Be fair and be consistent.

Twenty Recommendations

1) Under-promise. Over-perform:
- Repeat your intentions over and over, doing everything possible to prevent false expectations.
- Make sure you complete the task you promised.
- Build sustainable solutions that can be effectively transitioned to national agency management.
- Make sure it's legal to make the promises you choose to make:
    - Standards of Conduct and Rules of Engagement can be exceptionally complicated for all concerned;
    - Keep the lawyers within your inner circle.

2) The humanitarian intent should remain primary even against odds. Let it drive the operation.

3) Ensure all planning is Joint:
- Use your sister Services. Play to everyone's strengths.

4) Synergy between the civilian and military arenas can be found through an awareness of mutual advantage within the continuum of effort:
- Help each party see the valuable reasons for working together.
- Ensure they have those reasons.

5) A Civil-Military Operations Center, in function if not in physical space, will be indispensable:
- Break down the razor-wire barriers. Invite partners in. Choose neutral territory whenever possible. Avoid owning the space.
    - Co-location improves the cohesiveness of effort
    - Problems can be solved in a common forum
    - Issues discussed elsewhere can be voted upon collectively

6) Women in uniform can be a reassuring presence to an affected population:

- Remember that 85 percent of recent affected populations have been women and children.
- May distinguish the image of our new military presence over the local traditional military presence, increasing acceptance and improving rapport quickly.

7) Avoid compartmentalized planning
- Communication is hard enough. Don't compound the problem.

8) The two components likely to fail most frequently are communications and lift:
- Expect it. Plan in parallel layers.
- Keep a bedrock communications layer as a lowest common denominator across all boundaries and use it daily. Make it routine for everyone.

9) Readiness should not be confused with sustainability:
- Be ready for high-end force projection going in, with secure logistics to follow quickly. The sustainment force composition is very different, is rarely on tap, and must be accounted early.

10) The psychological costs of sustainment are disproportionately placed on the shoulders of the best talent. Protect them.

11) An afloat CMOC offers security, hygiene, and rest. While imperfect, it's a proven asset for some situations:
- Naval assets in a coastal region provide presence, poise, and protection. Bringing meetings aboard can be a welcome respite for all players, but it's not neutral and may be dangerous for those who must avoid misperceptions.
- In some circumstances, early, formative discussions between new partners (a Humanitarian Planning Team) can take place afloat, then move ashore when ready.

12) Critical Incident Stress Response within the care providers can prove disabling.
- Ensure resources are available for support. Most people have never seen what your personnel will have to endure daily.

13) Establish liaisons with stakeholders at every possible level, inserting full-time live bodies from coordinating agencies, such as the UN, USAID, senior NGOs, wherever it seems valuable and where you can:
- Within the Host Nation infrastructure
- Within the local population groups
- Within the coalition partners

14) Establish pre-conditions for deployment wherever you can within:
- Host nation agreements (through the Embassy, a Country Team, or a UN agency)
- International participants in the Theater
- UN agencies

- Donor nations and agencies

15) Establish a secure environment for the conduct of your mandate:
- Establish freedom of movement
- Neutralize the effectiveness of the belligerents in a fair and equitable manner across factions when possible
- Establish and maintain working relationships with the Host Nation, most frequently through the Embassy with a coordinating UN agency
- Be prepared to support humanitarian operations. They are often the core of the solution
- Be able to monitor, verify, and report on (and to) your major stakeholders

16) Information Operations can improve safety and security during transitions to sustainment:
- Conversely, in a digital age, communication dependent upon data links is inherently fragile and temptingly vulnerable. Expect failures. Plan multiple redundancies for critical paths.
- Use your PSYOPS and Civil Affairs teams early. May save lives.

17) Situational awareness will be problematic:
- Maps should be held in common and briefed in common. They are a critical resource and an opportunity for early rapport.
- Declassify information early and often to the greatest possible extent
- Daily briefs across topic areas should be held with all major stakeholders
    - Safety and Protection
    - Food
    - Logistics transport
    - Social services
    - Domestic needs
    - Health and nutrition
    - Water and sanitation
    - Education
    - Shelter
    - Income generation
    - Environmental protection
    - Agency operational support
    - Public Information
    - Budgets
    - Exit Strategy

18) Information that needs to be tracked constantly:
- Deployment of Armed Forces
- Stock at a glance
- Stock position, location
- Market price of food grains

- Maps: storing places and ports
- Shipping Schedules
- Position of Ships
- Unloading details
- Food movement programs
- Internal Procurement
- Maps: Situation Maps of affected areas (e.g. landmines...)
- Relief Activities
- Cash Allocated
- Damage Reports
- Foreign Relief arrivals

19) Key initial decisions to make:

i) Weapons status: When civilian and military are working together, consider reducing the personal weapons load out to the absolute minimum level required.
ii) Host Nation ability to provide an appropriate and adequate level of security
 (1) Try to augment and build. Not replace and take over.
iii) Medical support and social sensitivities
iv) Whether Civil Affairs teams are to do humanitarian assessments using tools held in common with UN relief agencies and NGOs
v) Means of communication coordination: Chose lowest common denominator. Include all possible parties.
vi) Imaging is necessary. Get the products to the people who really need it.
 (1) Again, declassify quickly
vii) Determine how to get information to the people who need the most support. Remember those 19 year olds in desert cammies on a street corner, implementing foreign policy with no strategic or operational view. Keep their educational needs paramount.

20) Know the environment

a) Weather, dust, mud, insects, diseases, and the impact each has on equipment and personnel

Thirty Advisories

1) Decide on the image you want to portray and stick to it.

2) Start and restart key institutions early
a) Medicine, education, water, telephone services, electrical power, churches
b) Begin the restoration of normalcy in areas that can transition early
c) Stay involved enough to ensure the re-start is sustainable

3) Don't make enemies but, if you do, don't treat them gently.

4) Encourage innovation and non-traditional approaches. Then listen when you get them.

5) Plan early and include everyone. This can't be stressed enough.

6) Determine the commander's intent, the centers of gravity, a mission analysis from the intent and those centers, an end state, measures of effectiveness toward that end state, and a phased exit strategy toward a transition to normalcy

7) Remain aware of other operational commitments elsewhere that may tax your resources and replenishment options. There have been unwelcome historical surprises.

8) Initial and replenishment manning will fall short in medical, dental, construction, port operations, and other specialist areas. Few services are deep in skilled personnel anywhere in this new millennium.

9) Avoid becoming the carrier of the United States checkbook. That can create a falsely inflated economy that is very hard on the rest of the local population.

10) Medical facilities have historically been looted and destroyed early.

11) Other skilled local staffs are often gone by the time we get there. Construction, electrical, plumbing, teachers, doctors, nurses, midwives, and so forth can be hard to find when you need them.

12) Be aware of the local anti-intervention PSYOPS campaign that may be waged against you.

13) Decide whether you need better maps and charts early and get the surveys started.
   a) Don't forget to share.

14) Recognize that fear of the military will be a constant impediment for many people at some level. It will be essential for you to understand the conflict issues, parties, and history.

15) Identify your budget and financial sponsors early, and recognize that all players present have both.

16) Also note that all participants have media requirements to one degree or another. Don't fight a tidal wave. Assist each other in the common goal of ensuring each is (and looks) useful to the sponsoring agencies.

17) Recognize that military priorities for early self-sustainment may not be shared by all parties present. Offload and airlift schedules will be controversial if transport logistics are difficult and controlled by the military.

18) Investigations of events that may later be evidence of war crimes will be an early requirement, often within the first 48 hours within the area. Be prepared to interview in the native language, photograph sites and subjects, and then archive the developing documentation. This has historically gone poorly and the guilty have gone free.

19) Be prepared to offer religious services to affected populations. In many areas of the world such services are an overriding concern. Hosting appropriate religious services can sometimes defuse significant tension.

20) There is never enough scalable amphibious lift in a coastal operation.

21) Make early arrangements for a mobile public address system and loan it to your partners when they need it.

22) Form an early RF communications link between the military, USAID, and the UN agencies.

23) There should be an area available for informal discussions with coffee, tea, and a few chairs. Much gets accomplished in an informal context, one-on-one, face-to-face.

24) The UN agencies need a separate area to work together apart from the military. Don't resent it; their safety is dependent upon impartiality, and they have no weapons. In addition, they will be there far longer than most military staff and may well stay in better, long-term billeting. Avoid jealousy. Avoid mocking. They know they'll be there long after we've gone.

25) Early formation of a Combined Logistics Cell that incorporates everyone saves a lot of time, money, and frustration. WFP is especially talented at logistics and is a useful resource.

26) Power fails. Have backup manual methods for anything important that requires electricity. There are many painful lessons on record.

27) Have a central area for posting decisions, messages, and schedules.

28) Remember that more than half of our recent populations have been under age 15. We need preparation to handle children in large numbers. Use UNICEF. They are prepared to manage many "vulnerable population" issues, including child soldiers.

29) The UN agencies are very reluctant to depend on the military for core services support. Our military priorities are altered on short notice from authority outside of our control, potentially affecting UN agency provisioning to an affected population with little warning and with no recourse. They won't rest easily if we're the sole source for something important.

30) Ensure all of your personnel understand their Standards of Conduct and Rules of Engagement. The carrying of small laminated cards has proved useful.

APPENDIX E

DOD/Military Documents, Field Manuals, Joint Publications

1. "Transforming for Stabilization and Reconstruction Operations," Hans Binnendijk and Stuart Johnson, National Defense University, 2004.
2. "DoD Net-Centric Data Strategy," ASD NII, May 9, 2003
3. DoD Instruction 8110.1, Multinational Information Sharing Networks Implementation, 6 February 2004
4. FM 100-23 Peace Operations
5. FM 100-6 Information Operations
6. FM 3-07 Stability Operations and Support Operations
7. FM 3-07.31 Peace Operations: Multi-Service Tactics, Techniques and Procedures for Conducting Peace Ops
8. FM 27-5, Manual of Military Government and Civil Affairs, 1940
9. Joint Publication 3-0, "Doctrine For Joint Operations"
10. Joint Publication 3-07, "Joint Doctrine for Military Operations Other Than War"
11. Joint Publication 3-07.3, "JTTP For Peace Operations"
12. Joint Publication 3-07.6, "JTTP for Humanitarian Assistance"
13. Joint Publication 3-08, "Interagency Coordination During Joint Operations"
14. Joint Publication 3-57, "Joint Doctrine for Civil-Military Operations"
15. Joint Publication 3-57.1, "Joint Doctrine for Civil Affairs"
16. Interagency Management of Complex Crisis Operations (Handbook), NDU, January 2003
17. Defense and Technology Paper: "Transforming the Reserve Component," NDU Center for Technology and national Security Policy
18. GTA 41-01-003, Civil Affairs Foreign Humanitarian Assistance Planning Guide, JQ department of the Army
19. Defense Science Board, "Report of the Defense Science Board 2004 Summer Study on Transition to and From Hostilities." (December 2004)

Other Agency Handbooks, Guidelines, & Strategies:

1. United Nations High Commissioner for Refugees *Handbook for Emergencies*
2. United Nations Civil-Military Coordination handbook
3. United Nations Disaster Assessment and Coordination Field Handbook
4. World Health Organization booklet *New Emergency Health Kit*
5. United Nations Children's Fund handbook entitled *Assisting in Emergencies*
6. U.S. State Department's Bureau for Population, Refugees, and Migration
7. *Assessment Manual for Refugee Emergencies*
8. U.S. Public Health Service *Handbook of Environmental Health*
9. World Food Programme Contingency Planning Guidelines
10. *Field Operations Guide (FOG) for Disaster Assessment and Response* Version 3.0, the U.S. Agency for International Development/Office of Foreign Disaster Assistance (OFDA)

11. Worldwide Humanitarian Assistance Logistics System (WHALS) Handbook, IDA
12. Handbook on Emergency Telecommunications, 2005, ITU
13. Guide to IGOs, NGOs and the Military in Peace and Relief Operations, Aall et.al., USIP
14. *NGO Coordination at Field Level: A Handbook*, John Bennett, ICVA
15. US Agency for International Development (USAID), Fragile States Strategy. (January 2005).
16. UN OCHA Civil-Military Coordination training material, 2005
17. Guidelines On The Use of Military and Civil Defence Assets To Support United Nations Humanitarian Activities in Complex Emergencies, March 2003

U.S. Government Agency, Multinational Organization, and Research Center Studies and Working Papers:

1. "Creating a Common Communications Culture", USIP, Virtual Diplomacy Initiative, 2004
2. "Winning the Peace: An American Strategy for Post-Conflict Reconstruction', edited by Robert C. Orr, The CSIS Press, Washington, D. C., 2004
3. "Multilateral Interoperability Programme: The C2 Information Exchange Data Model", Greding, Germany, October 1, 2004.
4. Miscellaneous S&R related briefing material, Martin Lidy, IDA, 2005
5. "Humanitarian Knowledge Management", Dennis King, U.S. Department of State, Humanitarian Information Unit, ISCRAM Conference, April 2005.
6. "Pre- and Post-Conflict Stability Operations," Joint DARPA and CSIS Workshop, June 2004
7. Hamre, John, "Civilian Post-Conflict Reconstruction Capabilities," CSIS Document, March 2003
8. Thie, Harry, et. al, "Framing a Strategy for Joint Officer Management: Preparing Officers for Interagency and Multinational Service," RAND Publication.
9. "Towards Interoperability in Crisis Management," USIP, September 2003
10. Amb. Pascual, Carlos (S/CRS), "Unifying Our Approach to Conflict Transformation," Remarks Assoc. of the US Army Conference, October 2005
11. "Building Civilian Conflict-Response Capabilities," S/CRS Document, May 2005
12. Hayes and Sands, "Doing Windows: Non-Traditional Military Responses to Complex Emergencies," DODCCRP Publication.
13. Robert M. Perito, *Where is the Lone Ranger When You Need Him?*, United States Institute of Peace, Washington, D.C., 2004
14. Hans Binnendijk and Stuart Johnson (eds.), "Transforming for Stabilization and Reconstruction Operations," Center for Technology and National Security Policy, National Defense University, Washington, D.C., 15 December 2003
15. William Flavin, *Civil-Military Operations: Afghanistan*, U.S. Army Peacekeeping and Stability Operations Institute, Carlisle, PA, Draft 3.2, 8 June 2004
16. Anthony H. Cordesman, *The "Post-Conflict" Lessons of Iraq and Afghanistan,* Center for Strategic and International Studies, 19 May 2004

17. Ray S. Jennings, *The Road Ahead: Lesson in Nation Building from Japan, Germany, and Afghanistan for Postwar Iraq*, U.S. Institute of Peace Peaceworks No. 49, Washington, D.C., April 2003

18. *Lessons from Bosnia: The IFOR* Experience, 1998, *Lessons from Kosovo: The KFOR Experience*, 2002, Larry Wentz (ed.), DoD Command and Control Research Program, Washington, D.C.

19. "America's Role in Nation-Building From Germany to Iraq," James Dobbins, RAND, 2003.

20. "Lessons from the Past: The American Record on Nation Building," Minxin Pei and Sara Kasper Carnegie Endowment for International Peace, 2003.

21. "Burden of Victory," James T. Quinlivan, RAND Review, Summer 2003

22. "Iraq's Post-Conflict Reconstruction: A Field Review and Recommendations," Dr John Hamre, CSIS, 2003

23. *Playing to Win* and *A Wiser Peace: An Action Strategy for A Post-Conflict Iraq*, CISI Report, January 2003

24. *Post-Conflict Reconstruction Task Framework*, CISI Report, 2002

25. "Winning the Peace: An American Strategy for Post-Conflict Reconstruction', edited by Robert C. Orr, The CSIS Press, Washington, D. C., 2004

26. "Engineering the Peace: The Military Role in Post-Conflict Reconstruction," Col Garland Williams, USIP

27. The Sphere Project: Humanitarian Charter and Minimum Standards in Disaster Response, First final edition 2000, Oxfam Publishing ISBN 0-85598-445-7

28. Report on Integrated Missions: Practical Perspectives and Recommendations Independent Study for the Expanded UN ECHA Core Group, May 2005

29. UN OCHA, Civil-Military Relationship in Complex Emergencies- An IASC Reference Paper -28 June 2004

30. UN OCHA, Use Of Military Or Armed Escorts For Humanitarian Convoys, 2001

31. AJP-9: NATO Civil-Military Co-Operation (CIMIC) Doctrine, 2001

32. Post-Conflict Reconstruction ESSENTIAL TASKS, Office of the Coordinator for Reconstruction and Stabilization Department of State, April 2005

33. "A Bridge Too Far: A id Agencies and th e Military in Hum anitarian Response," Barry, J. and Jefferys, A., Humanitarian Practices Network, January 2002

34. "Respect for Humanitarian Mandates in Conflict Situations," IASC Document, 1996

35. *On the Brink: Weak States and US National Security*, Center for Global Development Commission on Weak States and US National Security, June 2004

36. *Building Capacity for US Stability Operations: The Rule of Law Component*, USIP Special Report 118, April 2004

37. "Humanitarian Coordination: Lessons from Recent Field Experience," A study commissioned by the UN OCHA, June 2001

38. Gompert, David, et al, "Stretching the Network: Using Transformed Forces in Demanding Contingencies Other Than War," RAND Publication, 2004

Other Publications:

1. Barnett, Thomas, *The Pentagon's New Map: War and Peace in the 21st Century*, (New York : G.P. Putnam's Sons, 2004).
2. Francis Fukuyama, "Nation Building 101," *The Atlantic Monthly*, January/February 2004, available at http://www.theatlantic.com/doc/200401/fukuyama.
3. Fallows, James, "The Hollow Army," *The Atlantic Monthly*, March 2004, available at http://www.theatlantic.com/doc/200403/fallows.
4. Traub, James, "Making Sense of the Mission," *New York Times Magazine*, April 11, 2004.
5. Latham, Robert, editor, *Bombs and Bandwidth*, (New York: The New Press, 2003).
6. Phillips, David L., *Losing Iraq: Inside the Postwar Reconstruction Fiasco*, (Boulder: Westview Press, 2005).
7. Schultheis, B, *Waging Peace: A Special Operations Team's Battle to Rebuild Iraq*, (New York: Gotham Books, 2005).
8. Pascual, Carlos, and Krasner, Stephen. "Addressing State Failure," *Foreign Affairs*, vol. 84. no.4, 153. July/August 2005, 153-164
9. Covey, Jock, Dziedzic, Michael J., and Hawley, Leonard R., eds. *The Quest for Viable Peace: International Intervention and Strategies for Conflict Transformation*. U.S. Institute of Peace/Association of the U.S. Army, (Washington DC: United States Institute of Peace, 2005).
10. Collier, Paul, "Development and Conflict," Center for the Study of African Economies, Department of Economics, Oxford University, October 1, 2004